P9-DFI-051

Springer Series in

MATERIALS SCIENCE 132

Springer Series in
MATERIALS SCIENCE

The Springer Series in Materials Science covers the complete spectrum of materials physics, including fundamental principles, physical properties, materials theory and design. Recognizing the increasing importance of materials science in future device technologies, the book titles in this series reflect the state-of-the-art in understanding and controlling the structure and properties of all important classes of materials.

Please view available titles in *Springer Series in Materials Science* on series homepage http://www.springer.com/series/856

Ronald Redmer
Bastian Holst
Friedrich Hensel
Editors

Metal-to-Nonmetal Transitions

With 76 Figures

Editors
Prof. Dr. Ronald Redmer
Bastian Holst
University of Rostock
Institute of Physics
18051 Rostock, Germany
E-mail: ronald.redmer@uni-rostock.de
bastian.holst@uni-rostock.de

Prof. Dr. Dr. h.c. mult. Friedrich Hensel
Philipps-University Marburg
Department of Physical Chemistry
Hans-Meerwein-Straße
35032 Marburg, Germany
E-mail: hensel@mailer.uni-marburg.de

Springer Series in Materials Science ISSN 0933-033X
ISBN 978-3-642-03952-2 e-ISBN 978-3-642-03953-9
DOI 10.1007/978-3-642-03953-9
Springer Heidelberg Dordrecht London New York

Library of Congress Control Number: 2009943834

Cover design: SPi Publisher Services

Printed on acid-free paper

Springer is part of Springer Science+Business Media (www.springer.com)

Preface

Materials can be divided into metals and nonmetals. The characteristic feature of metals like copper and aluminum is a high electrical and thermal conductivity, while nonmetals such as phosphor and sulfur are insulators. The electrical conductivity varies over many orders of magnitude, from $10^6\,\Omega^{-1}\,\mathrm{m}^{-1}$ for typical metals down to $10^{-20}\,\Omega^{-1}\,\mathrm{m}^{-1}$ for almost ideal insulators. However, a sharp separation between metals and insulators is in general not possible. For instance, semiconducting materials such as silicon and germanium fill the conductivity domain between metals and insulators. Their electrical conductivity is dependent on temperature and, in addition, can be varied strongly by doping the material with donor or acceptor atoms. A famous example is the sharp insulator-to-metal transition measured in Si:P at temperatures below 0.1 K and donor concentrations of about $3.8 \times 10^{18}\,\mathrm{cm}^{-3}$ phosphor atoms [1].

Furthermore, materials may exist in both states: carbon is metallic as graphite and insulating as diamond. A fascinating quantum effect is observed at low temperatures: some materials even loose their electrical resistivity and become superconductors. Therefore, the questions *What is a metal?* and *When does a metal transform into a nonmetal?* are of fundamental interest and related to many aspects of modern physics and chemistry. We refer the interested reader to the very nice introduction into this diverse topic given by Edwards [2].

This book offers a collection of reviews on nonmetal-to-metal (or metal–insulators or Mott transitions) in very different physical systems, from solids with a regular periodic structure via disordered fluids and plasmas, finite metal clusters up to exotic nuclear and quark matter. The surprising similarity in the behaviour of these very diverse systems is due to the complex many-body nature of the respective interactions, which drives the transition and entails a non-perturbative treatment. Therefore, the Mott transition can be regarded as a prominent test case for methods of non-perturbative many-body physics. This book aims to give an overview on the current status of the theoretical treatment of Mott transitions and new experimental progress and findings in these fields as well.

In his original work, Mott [3] initiated the detailed discussion of metal–insulator transitions with an analysis of the critical screening length required to trap an electron around a positive ion in a solid from which he derived the relation $n^{1/3}a_\mathrm{B} = 0.2$. This famous Mott criterion has been very successful in describing metal–insulator transitions in various ordered systems, for example solids and doped semiconductors. Then Hubbard [4] introduced his today well-known and intensively studied model such that interactions between electrons are accounted for only when they are on the same site – via the repulsive Hubbard U term. Considering disordered systems, Anderson [5] could show that at a certain degree of disorder all electrons will be localized and the system becomes non-conducting.

These basic models contain important physical effects such as screening, repulsive on-site electron–electron interactions and disorder in a clear conceptual way and were, therefore, studied extensively. Details can be found in earlier reviews on this topic [6–9]. In real physical systems, we have to treat *all* relevant correlation and quantum effects to account for finite temperatures and thus thermal excitations and, where applicable, to include the influence of disorder as well. The simultaneous occurrence of correlations and disorder and their mutual interplay is of major importance in this context as has long been stressed by Mott. The construction and evaluation of respective Mott–Hubbard–Anderson-type models is one of the most challenging problems of many-body physics, see for example [10, 11].

The transition from a non-conducting to a conducting state in, for example electron–ion systems is connected with a change in the electronic wave function from being localized on a single atom or at few sites to a delocalized state. Landau and Zeldovich proposed already in 1943 that this electronic transition could introduce additional lines of first-order transitions in the phase diagram of the fluid state [12]. Their prediction has stimulated precise measurements of the liquid–vapour phase transition in metallic fluids such as mercury up to the critical point, see [13]. A new interpretation of data for the combined liquid–vapour and metal-to-nonmetal transition in mercury is given in Chap. 2. Furthermore, this electronic transition may have a strong impact on the high-pressure phase diagram of, for example hydrogen as the simplest and most abundant element [14, 15]. Extreme states of matter, that is pressures of several megabar and temperatures of many thousand Kelvin, occur in the interior of giant planets in our solar system as well as in extrasolar giant planets, which have been detected in great number. A better understanding of their formation processes, their current structure and evolution is intimately related to the high-pressure equation of state and the location of phase transition lines in fluid hydrogen–helium mixtures, see also Chap. 4.

It is obvious that the energy spectrum of electron–ion states, which contains in general a series of bound states at discrete, negative energies as well as a continuum of scattering states at positive energies, plays a central role for the understanding of the metal–nonmetal transition. The energy spectrum can be calculated by solving effective two-particle Schrödinger or

Bethe–Salpeter equations, which contain the correlation and quantum effects in a strongly coupled system via a perturbative treatment, or within improved self-consistent schemas such as the GW approximation, see [16–18]. In particular, the properties of *bound states* (formation, life time and dissolution) have been studied extensively in partially ionized plasmas as function of density and temperature.

Bound states in Coulomb systems are atoms (excitons) in electron–ion (electron–hole) plasmas and fluids. This concept can be generalized to nuclear matter where deuterons or alpha particles as found in neutron stars or in heavy ion collisions are bound states composed of nucleons that interact via effective nucleon–nucleon potentials. Bound states occur also in quark matter as diquark (e.g. pi-meson) or three-quark states (nucleons). The transition to a quark–gluon plasma can then be interpreted as the dissolution of all respective multi-quark states, similar to the transition from a partially to a fully ionized electron–ion or electron–hole plasma. Driving force is in all cases an increase in pressure or density. Thus, the original concept of Mott has found wide applications beyond traditional Coulomb systems, and the respective Mott transition is intensively studied.

In the following chapters, we present reviews on the Mott transition in these various systems, which will address the specific questions as well as the general problems. We start in Chap. 1 with a description of quantum phase transitions in strongly correlated one-dimensional electron–phonon systems and a detailed discussion of the models of Luttinger, Peierls and Mott. A new inspection of the metal–nonmetal transition in fluid mercury is given in Chap. 2, which has revealed a non-congruent nature for the first time. This might have consequences also for other predicted first-order phase transitions such as the hypothetical plasma phase transition in warm dense matter (see Chaps. 3 and 4), various phase separations in dusty plasmas, or the exotic phase transitions in neutron stars (see Chaps. 6 and 7). Various aspects of the Mott effect in dense fluids and plasmas have been treated up to now, but Pauli blocking as a direct quantum statistical effect is a novel topic and will be discussed in Chap. 3 within a chemical model. The metal–insulator transition in dense hydrogen is of primary importance for modeling interiors of Jupiter-like giant planets. A confrontation of advanced chemical models with quantum molecular dynamics simulations within a strict physical picture is performed in Chap. 4. The so far hypothetical plasma phase transition is discussed both in Chap. 3 and in Chap. 4. Metal–insulator transitions can also be induced in small metal clusters by irradiation with intense and short laser pulses. The highly effective energy deposition by resonance absorption, the various ionization processes (tunnel, field, impact) and the subsequent Coulomb-driven cluster explosion process are described in Chap. 5. The Mott effect in nuclear matter is reviewed in Chap. 6 within a cluster mean-field approximation. For instance, the formation of a two-nucleon quantum condensate is observed. The properties of the condensate are strongly influenced by the bound states immersed in the dense medium, that is by the Mott effect. A quantum field theory for the understanding of the phase diagram of exotic quark matter is outlined in Chap. 7. The crossover between Bose–Einstein condensation of diquark bound states and condensation of diquark resonances is discussed in close relation to the usual Mott effect.

At this point, we express our greatest respect to the enormous and pioneering work of Sir Nevil Mott. Without his outstanding contributions, our knowledge of fundamental interaction and correlation effects in various fields of physics would be much poorer today. His work has inspired many physicists worldwide, among them also theory groups in Germany, especially in Rostock, Greifswald and Berlin, who have developed new concepts based on Mott's ideas for the metal–insulator transition in fluids and plasmas as well as in nuclear and quark matter. Therefore, it was self-evident to celebrate Mott's 100th birthday on 30 September 2005 at the University of Rostock by dedicating an International Workshop to the subject of *Nonmetal–Insulator Transitions in Solids, Liquids and Plasmas*; participants of the meeting are shown in Fig. 1. The contributions to this book are mainly based on lectures given on that occasion or were invited afterwards:

Fig. 1. Participants of the International Workshop in Rostock on the occasion of Mott's 100th birthday on 30 September 2005 (from *left* to *right*): F. Hensel, B. Holst, D. Semkat, N. Nettelmann, A. Kietzmann, A. Kleibert, J. Adams, A. Bechler, M. French, T. Fennel, R. Egdell, K.-H. Meiwes-Broer, T. Döppner, W. Ebeling, J. Berdermann, T. Bornath, H. Reinholz, W.-D. Kraeft, V. Schwarz, H. Stolz, R. Ludwig, G. Röpke, R. Redmer, A. Weiße and D. Kremp.

- P.P. Edwards (Oxford): Phase Separation in Metal–Ammonia Solutions: Was Mott, or was Ogg Correct?
- R. Egdell (Oxford): Electron Spectroscopy and Metal-to-Nonmetal Transitions in Oxides
- H. Stolz (Rostock): Mott Effect and Bose–Einstein Condensation in Dense Exciton Systems
- G. Röpke (Rostock): Mott Effect in Nuclear Matter: Formation of Deuterons at Finite Temperature and Density
- W. Ebeling (Berlin): On Coulombic Phase Transitions
- F. Hensel (Marburg): Electronic Transitions in Liquid Metals
- K.-H. Meiwes-Broer (Rostock): Metal–Insulator Transitions in Expanding Clusters
- R. Redmer (Rostock): Metal–Nonmetal Transition in Dense Plasmas

We thank all contributors to this book for the careful preparation of their manuscripts. Finally, the project could be finished successfully and we thank all authors for their patience and for staying tuned to the project until the end. We thank Peter Edwards for his interest and the continuing support.

We thank the Deutsche Forschungsgemeinschaft (DFG) for support within the SFB 652, especially for the organization of the Workshop in 2005. Finally, we thank the Spinger-Verlag for supporting our project and, especially, Mr. Balamurugan Elumalai for the excellent mentoring of the edition of this book.

References

1. T.F. Rosenbaum, K. Andres, G.A. Thomas, R.N. Bhatt, Phys. Rev. Lett. **45**, 1723 (1980)
2. P.P. Edwards, in *The New Chemistry*, ed. by N. Hall, (Cambridge University Press, Cambridge, 2000), p. 85
3. N.F. Mott, Proc. Phys. Soc. **A62**, 416 (1949)
4. J. Hubbard, Proc. R. Soc. Lond. **A227**, 237 (1964); **281**, 401, (1964)
5. P.W. Anderson, Phys. Rev. **109**, 1492 (1958)
6. N.F. Mott, *Metal–Insulator Transitions* (Taylor and Francis, London, 1990)
7. P.P. Edwards, C.N.R. Rao (eds.), *Metal–Insulator Transitions Revisited* (Taylor and Francis, London, 1995)
8. D. Belitz, T.R. Kirkpatrick, Rev. Mod. Phys. **66**, 261 (1994)
9. M. Imada, A. Fujimori, Y. Tokura, Rev. Mod. Phys. **70**, 1039 (1998)
10. D.E. Logan, Y.H. Szczech, M.A. Tusch, in *Metal–Insulator Transitions Revisited*, ed. by P.P. Edwards, C.N.R. Rao (Taylor and Francis, London, 1995), p. 343
11. P.P. Edwards, R.L. Johnston, C.N.R. Rao, D.P. Tunstall, F. Hensel, Phil. Trans. R. Soc. Lond. A **356**, 5 (1998); P.P. Edwards, M.T.J. Lodge, F. Hensel, R. Redmer, Phil. Trans. Roy. Soc. A **368**, 941 (2010)
12. L.D. Landau, Ya.B. Zeldovich, Acta Phys.-Chim. USSR **18**, 194 (1943)
13. F. Hensel, W.W. Warren Jr., *Fluid Metals* (Princeton University Press, Princeton, 1999)

14. W.J. Nellis, Rep. Prog. Phys. **69**, 1479 (2006)
15. V.E. Fortov, Physics - Uspekhi **50**, 333 (2007)
16. W. Ebeling, W.D. Kraeft, D. Kremp, *Theory of Bound States and Ionization Equilibrium in Plasmas and Solids* (Akademie, Berlin, 1976)
17. W.-D. Kraeft, D. Kremp, W. Ebeling, G. Röpke, *Quantum Statistics of Charged Particle Systems* (Akademie, Berlin, 1986)
18. D. Kremp, M. Schlanges, W.-D. Kraeft, *Quantum Statistics of Nonideal Plasmas. Springer Series on Atomic, Optical and Plasma Physics, Band 25* (Springer, Berlin, 2005)

Rostock
February 2010

Ronald Redmer
Bastian Holst
Friedrich Hensel

Contents

List of Contributors

David Blaschke
Institute for Theoretical Physics
University of Wroclaw
50-204 Wroclaw, Poland
blaschke@ift.uni.wroc.pl

Werner Ebeling
Institut für Physik
Humboldt-Universität Berlin
12489 Berlin, Germany
ebeling@physik.hu-berlin.de

Holger Fehske
Institut für Physik
Ernst-Moritz-Arndt-Universität
Greifswald
17487 Greifswald, Germany
holger.fehske@physik.
uni-greifswald.de

Thomas Fennel
Institute für Physik
Universität Rostock
18051 Rostock, Germany
thomas.fennel@uni-rostock.de

Georg Hager
Regionales Rechenzentrum Erlangen
Universität Erlangen
91058 Erlangen, Germany
georg.hager@rrze.
uni-erlangen.de

Friedrich Hensel
Fachbereich Chemie
Physikalische Chemie
Philipps-Universität Marburg
35032 Marburg, Germany
hensel@mailer.uni-marburg.de

Bastian Holst
Institut für Physik
Universität Rostock
18051 Rostock, Germany
bastian.holst@uni-rostock.de

Karl–Heinz Meiwes–Broer
Institut für Physik
Universität Rostock
18051 Rostock, Germany
meiwes@physik.uni-rostock.de

Ronald Redmer
Institut für Physik
Universität Rostock
18051 Rostock, Germany
ronald.redmer@uni-rostock.de

Heidi Reinholz
Institut für Physik
Universität Rostock
18051 Rostock, Germany
heidi.reinholz@uni-rostock.de

Gerd Röpke
Institut für Physik
Universität Rostock
18051 Rostock, Germany
gerd.roepke@uni-rostock.de

Joseph Tiggesbäumker
Institut für Physik
Universität Rostock
18051 Rostock, Germany
josef.tiggesbaeumker@
uni-rostock.de

Nguyen Xuan Truong
Institut für Physik
Universität Rostock
18051 Rostock, Germany
xuan.nguyen@uni-rostock.de

Daniel Zablocki
Institute for Theoretical Physics
University of Wroclaw
50-204 Wroclaw, Poland
zablocki@ift.uni.wroc.pl

1

Luttinger, Peierls or Mott? Quantum Phase Transitions in Strongly Correlated 1D Electron–Phonon Systems

Holger Fehske and Georg Hager

Abstract. We analyse the complex interplay of charge, spin, and lattice degrees of freedom in one–dimensional electron systems coupled to quantum phonons. To this end, we study generic model Hamiltonians, such as the Holstein models of spinless fermions, the Holstein–Hubbard model and a Heisenberg spin-chain model with magneto-elastic interaction, by means of an unbiased numerical density–matrix renormalisation group technique. Thereby particular emphasis is placed on the Luttinger–liquid charge–density-wave, Peierls–insulator Mott-insulator, and spin–Peierls quantum phase transitions.

1.1 Introduction

The way a material evolves from a metallic to an insulating state is one of the most fundamental problems in solid state physics. Apart from band structure and disorder effects, electron–electron and electron–phonon interactions are the driving forces behind metal–insulator transitions in the majority of cases. While the so-called Mott–Hubbard transition [1] is caused by strong Coulomb correlations, the Peierls transition [2] is triggered by the coupling to vibrational excitations of the crystal. Both scenarios compete in a subtle way. As a result, quantum phase transitions (QPT) between insulating phases become possible. Most notably this applies to quasi one-dimensional (1D) materials like conjugated polymers, organic charge transfer salts, ferroelectric perovskites, or halogen-bridged transition metal complexes, which exhibit a remarkably wide range of strengths of competing forces [3, 4]. Moreover, 1D systems are known to be very susceptible to structural distortions.

The challenge of understanding such a kind of metal–insulator or insulator–insulator QPT has stimulated intense work on generic microscopic models of interacting electrons and phonons. In this respect, the 1D Holstein–Hubbard model is particularly rewarding to study [5–10]. It accounts for a tight-binding electron band, a local coupling of the charge carriers to optical phonons, the energy of the phonon subsystem in harmonic approximation, and an intra-site Coulomb repulsion between electrons of opposite spin:

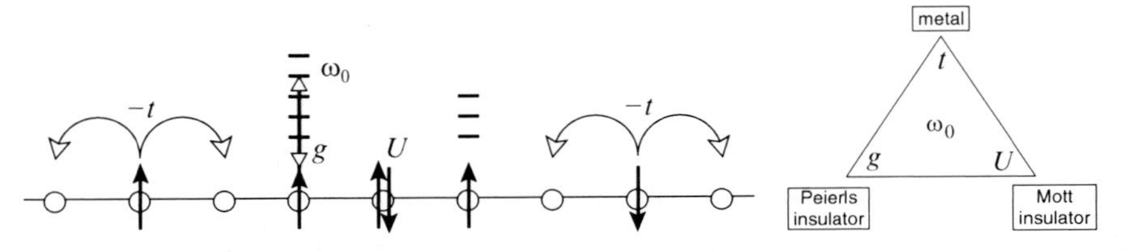

Fig. 1.1. The 1D Holstein–Hubbard model (*left panel*) and the competing ground states for the half-filled band case (*right panel*)

$$H = -t \sum_{\langle i,j \rangle \sigma} c_{i\sigma}^\dagger c_{j\sigma} - g\omega_0 \sum_{i\sigma} (b_i^\dagger + b_i) n_{i\sigma} + \omega_0 \sum_i b_i^\dagger b_i + U \sum_i n_{i\uparrow} n_{i\downarrow} \,. \quad (1.1)$$

Here $n_{i\sigma} = c_{i\sigma}^\dagger c_{i\sigma}$, where $c_{i\sigma}^\dagger$ ($c_{i\sigma}$) creates (annihilates) a spin-σ electron at Wannier site i of a 1D lattice with N sites, and $b_i^\dagger$ (b_i) are the corresponding bosonic operators for a dispersionsless phonon with frequency ω_0.

The physics of the Holstein[1]–Hubbard[2] model is governed by three competing effects: the itinerancy of the electrons ($\propto t$), their on-site Coulomb repulsion ($\propto U$), and the local electron–phonon (EP) coupling ($\propto g$). As the EP interaction is retarded, the phonon frequency (ω_0) defines a further relevant energy scale (see Fig. 1.1). This advises us to introduce besides the adiabaticity ratio, ω_0/t, two dimensionless coupling constants

$$u = U/4t \quad \text{and} \quad g^2 = \varepsilon_\text{p}/\omega_0 \quad \text{or} \quad \lambda = \varepsilon_\text{p}/2t \,. \quad (1.2)$$

Both Holstein and Hubbard interactions tend to immobilise the charge carriers. Therefore, Peierls insulator (PI) or Mott insulator (MI) states are expected to be favoured over the metallic state, at least for the half-filled band case ($\sum_{i,\sigma} n_{i\sigma} = N_\text{el} = N$) and at zero temperature. Strictly speaking, this holds in the adiabatic limit ($\omega_0 = 0$) for 'U-only' (Hubbard model) and 'λ-only' (Peierls model) parameters. For the more general Holstein–Hubbard model, the situation is much less obvious. Clearly a large phonon frequency will act against any static ordering. If insulating phases exist nevertheless, their ground-state properties will depend on ω_0 and on the ratio of Coulomb and EP interactions u/λ. Likewise, the nature of the physical excitations is puzzling as well. While one expects 'normal' electron–hole excitations in the PI phase ($U = 0$), charge (spin) excitations are known to be massive (gapless) in the MI state of the Hubbard model ($\lambda = 0$). Thus, varying the

[1] The Holstein model [11] has been studied extensively as a paradigmatic model for polaron formation in the low-density limit. For commensurate band fillings, the coupling to the lattice supports charge ordering.

[2] The Hubbard model [12], originally designed to describe ferromagnetism of transition metals, has more recently been used as the probably most simple model to account for strong Coulomb correlation effects in the context of high-temperature superconductivity.

control parameter u/λ, a cross-over from standard quasi-particle behaviour to spin-charge separation might be observed in the more general 1D Holstein–Hubbard model.

The aim of this contribution is to affirm this physical picture and the anticipated phase diagram of the 1D Holstein–Hubbard model. For these purposes we adapt Lanczos exact diagonalisation (ED) [13], kernel polynomial (KPM) [14] and density-matrix renormalisation group (DMRG) [15] methods for EP problems (for an overview see [16, 17]). These numerical techniques allow us to obtain unbiased results for all interaction strengths with the full quantum dynamics of phonons taken into account.

1.2 Luttinger–Peierls Metal–Insulator Transition

To study the metal–insulator transition in 1D EP systems, we neglect, in a first step, the spin degrees of freedom in (1.1). Even so, the resulting 1D spinless fermion Holstein model,

$$H = -t \sum_{\langle i,j \rangle} c_i^\dagger c_j - g\omega_0 \sum_i (b_i^\dagger + b_i) n_i + \omega_0 \sum_i b_i^\dagger b_i, \tag{1.3}$$

is, despite its seeming simplicity, not exactly solvable. It is generally accepted, however, that the model exhibits a QPT from a metallic to an insulating phase at half-filling ($N_\mathrm{e} = N/2$) [18,19]. During the last two decades, a wide range of analytical and numerical methods have been applied to map out the ground-state phase diagram in the whole $g-\omega_0$ plane [18, 20–26], with significant differences, especially in the low-frequency intermediate EP coupling regime. In the adiabatic limit ($\omega_0 \to 0$), the critical coupling $\lambda_\mathrm{c}(\omega_0)$ vanishes. In the anti-adiabatic ($\omega_0 \to \infty$) strong EP coupling regime, the model can be transformed to the exactly solvable XXZ model [18, 23], which shows a transition of Kosterlitz–Thouless type.

Before we determine the metal–insulator phase boundary, let us characterise the metallic and insulating phases themselves. According to Haldane's Luttinger liquid (LL) conjecture [27], an 1D gapless (metallic) system of interacting fermions should belong to the Tomonaga–Luttinger universality class [28,29]. As the Holstein model of spinless fermions is expected to be gapless at weak couplings g, the system is described by (non-universal) LL parameters u_ρ (charge velocity) and K_ρ (correlation exponent).

In the following, we try to determine u_ρ and K_ρ by large-scale DMRG calculations. To leading order, the charge velocity and the correlation exponent is related to the ground-state energy of a finite system with N sites

$$\frac{E_0(N)}{N} = \varepsilon_0(\infty) - \frac{\pi}{3} \frac{u_\rho}{2} \frac{1}{N^2} \tag{1.4}$$

($\varepsilon_0(\infty)$ denotes the bulk ground-state energy density) and the charge excitation gap

$$\varDelta_{\rm c}(N) = E_0^{\pm}(N) - E_0(N) = \pi \frac{u_\rho}{2} \frac{1}{K_\rho} \frac{1}{N} \tag{1.5}$$

(here $E_0^{\pm}(N)$ is the ground-state energy with ± 1 fermion away from half-filling $n = N_{\rm el}/N = 0.5$). Note that the LL scaling relations (1.4) and (1.5) were derived for the pure electronic spinless fermion model only [30]. A careful finite-size analysis shows, however, that they also hold for the case that a finite EP is included [31]. Figure 1.2 shows the resulting LL parameters, exemplarily for two frequencies belonging to the adiabatic (upper left panel) and anti-adiabatic (upper right panel) regimes. Interestingly, the LL phase splits into two different regions: for small phonon frequencies, the effective fermion–fermion interaction is attractive ($K_\rho > 0$), while it is repulsive ($K_\rho < 0$) for large frequencies. In the latter region, the kinetic energy ($\propto u_\rho$) is strongly reduced and the charge carriers behave like (small) polarons. In between, there is a transition line $K_\rho = 1$, where the LL is made up of (almost) non-interacting particles. The LL scaling breaks down just at a critical coupling $g_{\rm c}(\omega_0/t)$, signalling the transition to the CDW (charge density wave) state. We find, for example $g_{\rm c}^2(\omega_0/t = 0.1) \simeq 7.84$ and $g_{\rm c}^2(\omega_0/t = 10) \simeq 4.41$.

The middle panels of Fig. 1.2 prove the existence of CDW long-range order above $g_{\rm c}$. Here the staggered charge structure factor

$$S_{\rm c}(\pi) = \frac{1}{N^2} \sum_{i,j} (-1)^j \langle (n_i - n)(n_{i+j} - n) \rangle \tag{1.6}$$

unambiguously scales to a finite value in the thermodynamic limit ($N \to \infty$). Simultaneously, $\varDelta_{\rm c}(\infty)$ acquires a finite value. In contrast, we have $S_{\rm c}(\pi) \to 0$ in the metallic regime ($g < g_{\rm c}$). Note that such a finite-size scaling, including dynamical phonons, is definitely out of range for any ED calculation. The CDW at strong EP coupling is connected to a Peierls distortion of the lattice and can be classified as traditional band insulator and polaronic superlattice in the strong-coupling adiabatic and anti-adiabatic regimes, respectively.

The optical absorption spectra shown in the lower panels of Fig. 1.2 elucidate the different nature of the CDW for small and large adiabaticity ratios in more detail. The regular part of the optical conductivity,[3]

$$\sigma^{\rm reg}(\omega) = \sum_{m>0} \frac{|\langle \psi_0 | \hat{j} | \psi_m \rangle|^2}{E_m - E_0} \, \delta[\omega - (E_m - E_0)], \tag{1.7}$$

takes into account finite-frequency transitions from the ground state $|\psi_0\rangle$ to excited quasi-particle states $|\psi_m\rangle$ in the same particle sector.[4] Importantly, the current operator $\hat{j} = {\rm i}t \sum_i (c_i^\dagger c_{i+1} - c_{i+1}^\dagger c_i)$ has finite matrix elements between

[3] The evaluation of dynamical correlation functions like $\sigma^{\rm reg}(\omega)$ can be carried out by means of the very efficient and numerically stable ED-KPM algorithm [14].

[4] In (1.7), $\sigma^{\rm reg}(\omega)$ is given in units of πe^2 and we have omitted an $1/N$ prefactor.

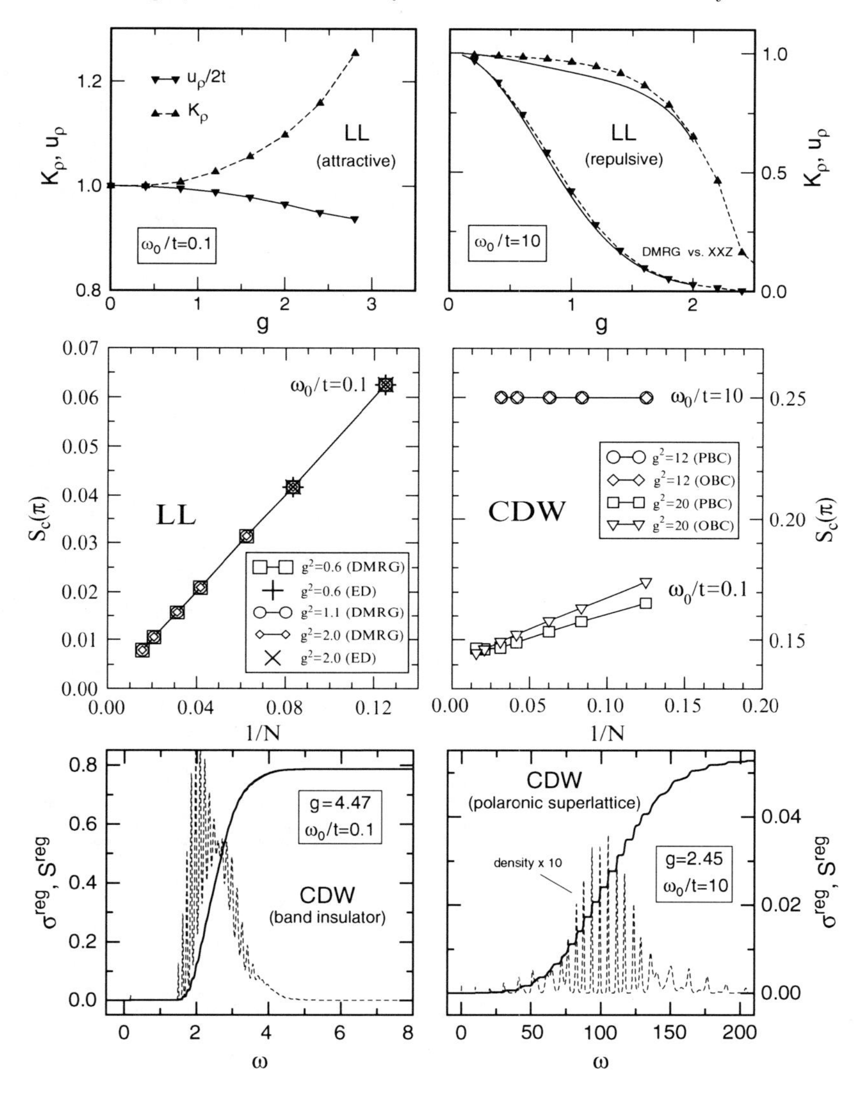

Fig. 1.2. Basic properties of the 1D half-filled spinless fermion Holstein model: Luttinger liquid parameters u_ρ and K_ρ in the metallic region (*top panels*; the *solid lines* in the right panel gives the asymptotic results for the XXZ model), finite-size scaling of the charge structure factor $S_c(\pi)$ below and above the metal–insulator transition (*middle panels*), and optical response $\sigma^{\text{reg}}(\omega)$ in the CDW regime (*lower panels*). See text for further explanation

states of different site-parity only. In the adiabatic region, the most striking feature is the sharp absorption threshold and large spectral weight contained in the incoherent part of optical conductivity. In the anti-adiabatic regime, the CDW is basically a state of alternate self-trapped polarons, which means that the electrons are heavily dressed by phonons. As the renormalised band dispersion is extremely narrow, finite-size gaps are reduced as well. Therefore, $\Delta_{\rm opt}$ read off from Fig. 1.2 yields the correct CDW gap.

Further information can be obtained from single-particle excitation spectra. The $T = 0$ electron spectral function is related to the one-electron Green function via

$$A(k,\omega) = -\frac{1}{\pi}\mathrm{Im}\, G(k,\omega) = A^{+}(k,\omega) + A^{-}(k,\omega)\,, \tag{1.8}$$

where

$$A^{\pm}(k,\omega) = -\frac{1}{\pi}\mathrm{Im}\lim_{\eta\to 0^{+}} \left\langle \psi_0 | c_k^{\mp} \frac{1}{\omega + \mathrm{i}\eta \mp H} c_k^{\pm} | \psi_0 \right\rangle\,, \tag{1.9}$$

with $c_k^- = c_k$, $c_k^+ = c_k^\dagger$. $A^-(k,\omega)$ [$A^+(k,\omega)$] describes [inverse] photoemission of an [injected] electron with momentum k and energy ω. The spectral functions shown in Fig. 1.3 have been calculated by an elaborate dynamical DMRG method [32, 33]. As we are in the insulating CDW phase, we observe a single-particle excitation gap at the Fermi momenta $k_{\rm F} = \pm\pi/2$. Below and above the gap, the spectrum shows broad multi-phonon absorption bands

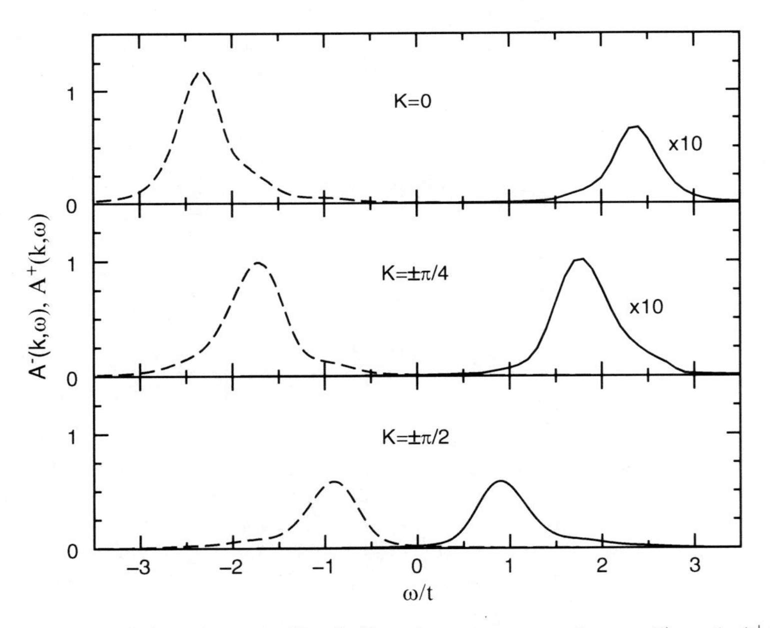

Fig. 1.3. Spectral functions $A^-(k,\omega)$ (for electron removal, $\omega < 0$) and $A^+(k,\omega)$ (for electron injection $\omega > 0$) for the spinless Holstein model at half-filling ($N = 10$ with periodic boundary conditions (PBC)). The system is in the CDW (Peierls insulating) phase ($g = 4$, $\omega_0/t = 0.1$). Note that only $|k| \le \pi/2$ is shown because $A(k \pm \pi,\omega) = A(k,-\omega)$

whose maxima roughly follow a renormalised cosine tight-binding dispersion in momentum space.

The $T = 0$ phonon spectral function is defined as

$$B(q,\omega) = -\frac{1}{\pi}\mathrm{Im}D(q,\omega), \tag{1.10}$$

with

$$D(q,\omega) = \lim_{\eta\to 0^+} \left\langle \psi_0|\hat{x}_q \frac{1}{\omega + \mathrm{i}\eta - H}\hat{x}_{-q}|\psi_0 \right\rangle \tag{1.11}$$

for $\omega \geq 0$ and $\hat{x}_q = N^{-1/2}\sum_j \hat{x}_j \mathrm{e}^{-\mathrm{i}(jq)}$. For the spinless fermion Holstein model (1.3), $B(q,\omega)$ is symmetric in q, and we have a dispersionsless bare propagator $D_0(q,\omega) = 2\omega_0/(\omega^2 - \omega_0^2)$. EP interaction will renormalise the phonon frequency, whereby $D(q,\omega)$ attains a q-dependence. To determine the q-dependence of $B(q,\omega)$ for the infinite lattice, we exploit ED in combination with cluster perturbation theory [26, 34]. The density plots shown in Fig. 1.4

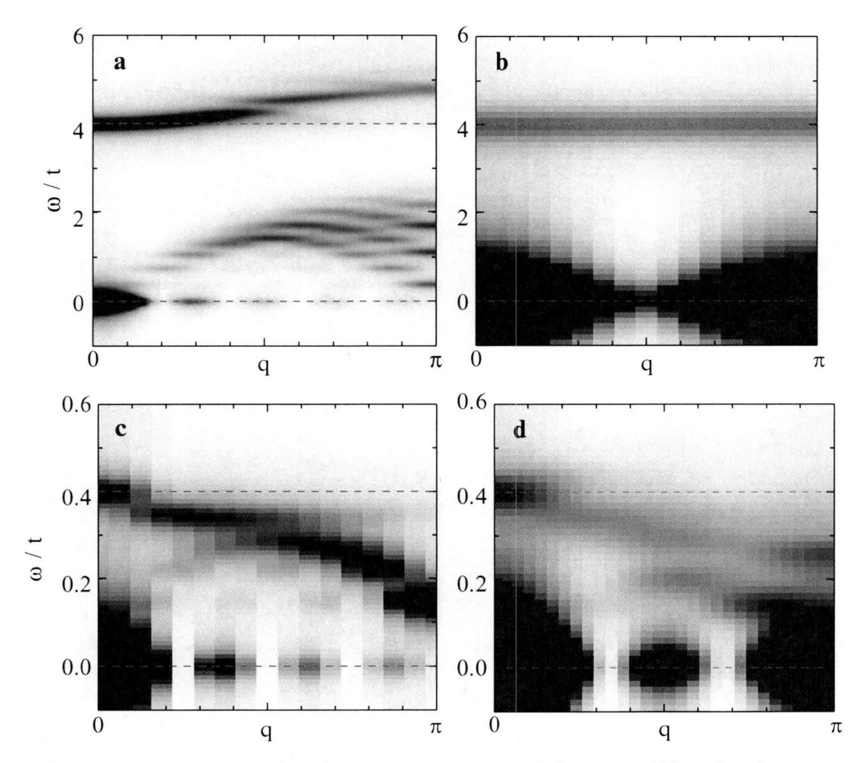

Fig. 1.4. Density plots of the CPT phonon spectral function $B(q,\omega)$, where panels (**a**)–(**d**) correspond to the repulsive LL ($\omega_0/t = 4$, $g^2 = 0.5$), polaron superlattice ($\omega_0/t = 4$, $g^2 = 5$), attractive LL ($\omega_0/t = 0.4$, $g^2 = 2.5$), and band insulator ($\omega_0/t = 0.4$, $g^2 = 5$) regimes, respectively (taken from [26])

summarise the differences between the anti-adiabatic and adiabatic regimes, and between the LL and CDW phases of the spinless fermion Holstein model. In the anti-adiabatic case, we observe two phonon signatures for all $g > 0$. In the LL phase, the bare phonon mode hardens, whereas a second mode becomes strongly over-damped near $q = \pi$ [panel (a)]. Panel (b) reveals a dispersionless signal at $\omega = \omega_0$, as well as the flat polaron band at $\omega \approx 0$ for the polaronic CDW state. Quite differently, in the adiabatic case [panel (c)], we see that renormalised phonon dispersion $\omega(q)$ softens with increasing EP coupling, leading to a degeneracy of excitations with $q = 0, \pi$ at g_c. Above the Peierls transition we find – in agreement with recent Monte Carlo simulations [35] – that the soft $q = \pi$ phonon mode splits into two branches with the upper one hardening as the EP coupling increases further [panel (d)].[5] Thus, with increasing phonon frequency, we find a cross-over from a soft-mode (displacive) to a central-peak-like (order-disorder-type) phase transition, similar to the analysis of the spin–Peierls transition motivated by $CuGeO_3$ [21] (cf. Sect. 5.1).

Finally the phase diagram of the 1D Holstein model is presented in Fig. 1.5. The regions where the system typifies as repulsive LL, attractive LL, polaronic superlattice, or band insulator are indicated. While we observe a rather smooth cross-over between different states within the metallic respectively in-

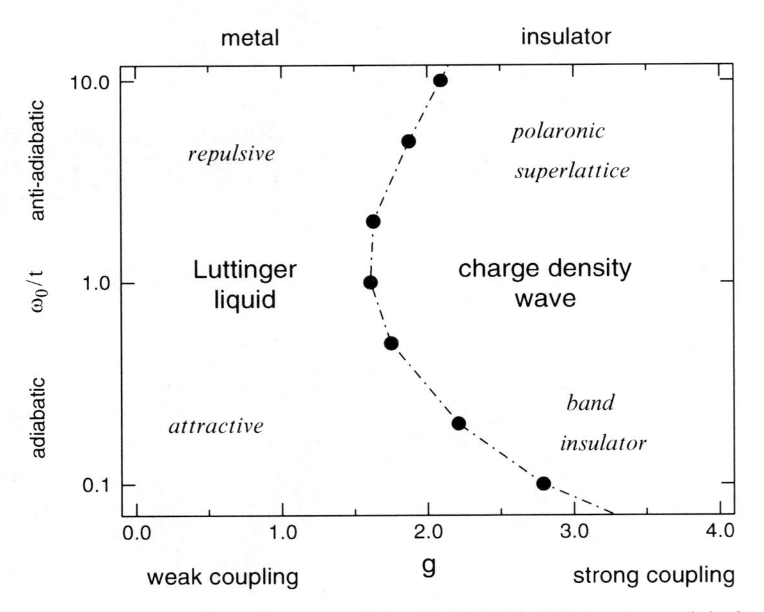

Fig. 1.5. Ground-state phase diagram of the 1D half-filled Holstein model of spinless fermions obtained by DMRG

[5] The strong zero-energy absorption feature at $\pi/2$ is an artifact of the small cluster size and the open boundary conditions used in the ED-CPT scheme.

sulating 'domains', there is a true metal–insulator QPT between the LL and the CDW phase. Three different methods were used to determine the phase boundary. First we employ an optimised phonon diagonalisation method [24] and look for an upturn of the charge structure factor in passing from LL to CDW. The second DMRG-based method was inspired by work on the frustrated Heisenberg model and exploits a level crossing criteria between the charge gap and the 'one-photon' excitation gap [23]. The third method rests on a DMRG finite-size scaling of the charge structure factor (cf. Fig. 1.2). The results basically agree and above all confirm that a finite critical EP coupling is required to set up the CDW phase provided $\omega_0 > 0$.

1.3 Peierls–Mott Insulator–Insulator Transition

Now we include the spin degrees of freedom and ask for the effect of a finite Coulomb interaction. The ground state of the pure Holstein model ($U = 0$) is a Peierls distorted state with staggered charge order for $g > g_c(\omega_0)$ [36,37], that is as in the Holstein model of spinless fermions, quantum phonon fluctuations destroy the Peierls state for $g < g_c$.

The charge structure factor, $S_c(\pi)$ [cf. (1.6)], and spin structure factor,

$$S_s(\pi) = \frac{1}{N^2}\sum_{i,j}(-1)^j\langle S_i^z S_{i+j}^z\rangle \quad \text{with} \quad S_i^z = \frac{1}{2}(n_{i\uparrow} - n_{i\downarrow})\,, \tag{1.12}$$

shown in Fig. 1.6 for the full Holstein–Hubbard model, indicate pronounced CDW and weak SDW (spin density wave) correlations, provided $u/\lambda < 1$. Increasing the Hubbard interaction u at fixed EP coupling λ and frequency ω_0, the CDW correlations become strongly suppressed, whereas the spin structure factor at $q = \pi$ is enhanced. To conclude about a possible existence of charge and/or spin long-range order, one has to determine $S_c(\pi)$ and $S_s(\pi)$ for different system sizes, followed by a finite-size scaling of the data. The size-dependence of the DMRG results for $S_{c/s}(\pi)$ is shown in the lower panels of Fig. 1.6. In the PI phase, $S_c(\pi)$ is almost constant and scales to a finite value, indicating true CDW long-range order, whereas $S_s(\pi)$ obviously scales to zero as $N \to \infty$. By contrast, in the MI regime, our data provides strong evidence for vanishing charge and also spin order in the thermodynamic limit. Consequently, the PI exhibits CDW order, that is alternating empty and doubly occupied sites, while the MI is characterised by short-ranged anti-ferromagnetic spin correlations (see Fig. 1.7).[6]

Figure 1.8 displays the (inverse) photoemission spectra for the Holstein–Hubbard model, where

$$A_\sigma^{\pm}(k,\omega) = \sum_m |\langle\psi_m^{\pm}|c_{k\sigma}^{\pm}|\psi_0\rangle|^2\,\delta[\omega \mp (E_m^{\pm} - E_0)], \tag{1.13}$$

[6] Here the staggered spin–spin correlations decay algebraically at large distances.

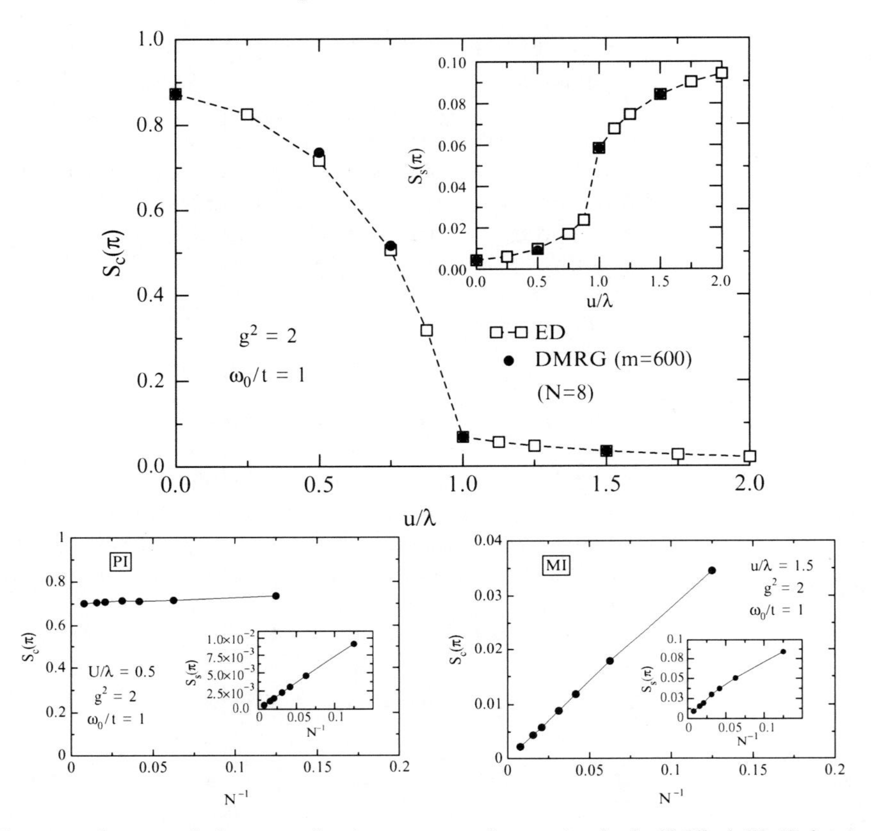

Fig. 1.6. Staggered charge and spin structure factors in the half-filled 1D Holstein–Hubbard model (1.1). The *upper panel* compares ED (*squares*) and DMRG (*filled circles*) results for a small eight-site system (PBC) with $\lambda = 1$, $\omega_0/t = 1$. The *lower panels* show the finite-size scaling of $S_c(\pi)$ and $S_s(\pi)$ (inset) in the PI (*left panel*) and MI (*right panel*) regimes

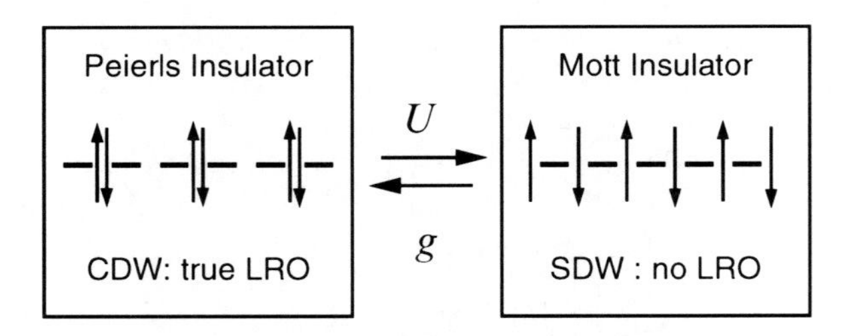

Fig. 1.7. Schematic structure of Peierls and Mott insulating states

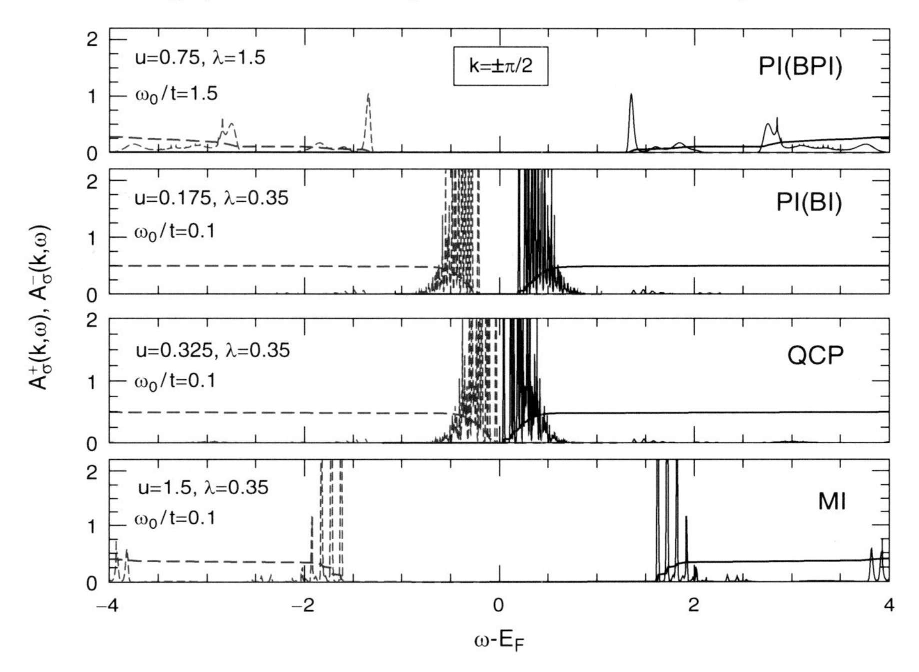

Fig. 1.8. Spectral densities for photoemission ($A_\sigma^-(k,\omega)$; *dashed lines*) and inverse photoemission ($A_\sigma^+(k,\omega)$; *solid lines*) in the $k = \pm\pi/2$ sector of the half-filled Holstein–Hubbard model. Data were obtained by applying our ED-KPM scheme. Shown are typical results obtained for the case of a bipolaronic insulator (BPI), a Peierls band insulator (BI), a system near the PI-MI quantum critical point (QCP), and a Mott insulator (MI) (from *top* to *bottom*). The corresponding integrated densities $S_\sigma^\pm(k,\omega) = \int_{\mp\infty}^{\omega} d\omega' A_\sigma^\pm(k,\omega')$ are also indicated

with $c_{k\sigma}^+ = c_{k\sigma}^\dagger$ and $c_{k\sigma}^- = c_{k\sigma}$ [cf. (1.9)]. To monitor a band splitting induced by the Hubbard and Holstein couplings, we focus on the results at the Fermi momenta $k_\mathrm{F} = \pm\pi/2$ (PBC).[7]

The most prominent feature we observe in the PI regime is a finite gap at $k = \pm\pi/2$. At high phonon frequencies, the insulating behaviour is associated with localised bipolarons forming a CDW state (BPI, upper panel). Because of strong polaronic effects, an almost flat band dispersion with exponentially small (electronic) quasi-particle weight results [6]. The dominant peaks in the incoherent part of the (inverse) photoemission spectra are related to multiples of the (large) bare phonon frequency. The situation changes if the phonon frequency is small (adiabatic regime). Here, for the traditional BI, a rather broad photoemission signature appears. Within these excitation bands, the spectral weight is almost uniformly distributed, which is a clear indication of the multi-phonon absorption and emission processes that accompany every

[7] Spectra for the other allowed momenta of our eight-site system are given in [6].

single-particle excitation in the PI. The line shape reflects the Poisson-like distribution of the phonons in the ground state. Away from the Fermi momenta, the lower and upper bands closely follow a (slightly renormalised) cosine dispersion. If we enhance the Hubbard interaction at fixed EP coupling strength, the Peierls gap weakens and finally closes at about $(u/\lambda)_{\rm c} \simeq 1$, which marks the PI-MI cross-over quantum critical point (QCP). This is the situation shown in the middle panel. Approaching the QCP, the ground state and the first excited state become degenerate. The QCP is characterised by gapless charge excitations at the Fermi momenta but perhaps should not be considered as metallic because the Drude weight in the case of a degenerate ground state is ill-defined [38]. If the Hubbard interaction further increases, Coulomb repulsion overcompensates the attractive on-site EP coupling and the single-particle excitation spectrum becomes gapped again (lowest panel). The Mott–Hubbard correlation gap almost coincides with the optical gap $\Delta_{\rm opt}$ determined by evaluating the regular part of the optical conductivity for the same parameters. The form of the MI spectra is quite different from PI case. Contrary to the PI phase in the MI regime, the lowest peak in each k sector is clearly the dominant one. The dispersion of the lower (upper) Hubbard band can be derived tracing the uppermost (lowest) excitations in each k sector. As N goes to infinity for $u \gg 1$, the lower Hubbard band will be completely filled, and consequently the system behaves as an insulator at $T = 0$.

The many-body charge and spin excitation gaps,

$$\Delta_{\rm c} = E_0^+(1/2) + E_0^-(-1/2) - 2E_0(0) \tag{1.14}$$

$$\Delta_{\rm s} = E_0(1) - E_0(0), \tag{1.15}$$

can also be used to characterise the different phases of the Holstein–Hubbard model. Here $E_0^{(\pm)}(S^z)$ is the ground-state energy at half-filling (with $N_{\rm e} = N \pm 1$ particles in the sector with total spin-z component S^z). As we compare ground-state energies calculating the charge and spin gaps, lattice relaxation effects arising from different particle numbers are included. This is of course not the case when determining the single-particle functions (1.9) or (1.13). Obviously, the DMRG finite-size scaling presented in Fig. 1.9 for $\Delta_{\rm c/s}$ substantiates our introductory discussion. $\Delta_{\rm c}$ and $\Delta_{\rm s}$ are finite in the PI and will converge further as $N \to \infty$. Compared to the BI phase, the finite-size dependence of $\Delta_{\rm c}$ and $\Delta_{\rm s}$ is much weaker in the BPI phase because the small bipolarons that emerge are rather localised objects. Both gaps seem to vanish at the quantum phase transition point of the Holstein–Hubbard model with finite-frequency phonons, but in the critical region the finite-size scaling is extremely delicate. In the MI we found a finite charge excitation gap, which in the limit $u/\lambda \gg 1$ scales to the optical gap of the Hubbard model, whereas the extrapolated spin gap remains zero. This can be taken as a clear indication for spin-charge separation.

So far we can summarise our findings by the schematic ground-state phase diagram shown in Fig. 1.10.

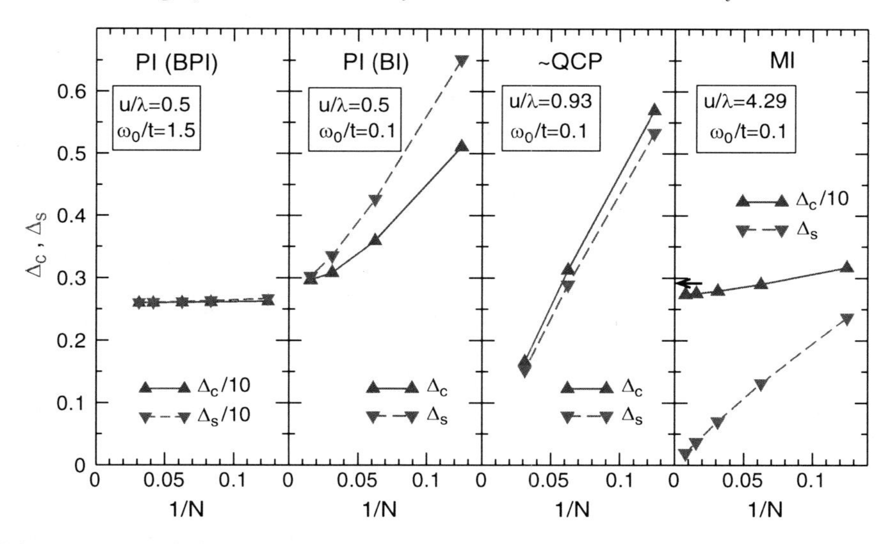

Fig. 1.9. DMRG finite-size scaling of spin- and charge-excitation gaps in the HHM at $\lambda = 0.35$ and $\omega_0/t = 0.1$, where open boundary conditions were used. The accessible system sizes are smaller at larger λ/u, where an increasing number of (phononic) pseudo-sites is required to reach convergence with respect to the phonons. The arrow marks the value of the optical gap $\Delta_{\rm opt}$ for the Bethe ansatz solvable 1D Hubbard model, which is given by $\Delta_{\rm opt}/4t = u - 1 + \ln(2)/2u$ in the limit of large $u > 1$ [39]

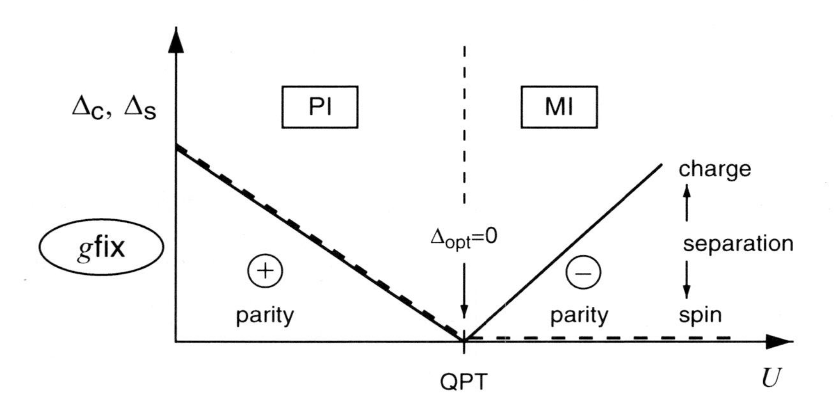

Fig. 1.10. PI-MI QPT in the strong coupling regime of the Holstein–Hubbard model. For finite periodic chains with $N = 4n$, the MI-PI quantum phase transition could be identified by a ground-state level crossing associated with a change in the parity eigenvalue $P = \pm$ [40], where the site inversion symmetry operator P is defined by $Pc_{i\sigma}^{\dagger}P^{\dagger} = c_{N-i\sigma}^{\dagger}$ [41].

1.4 On the Possibility of an Intervening Metallic Phase

Recently there has been some speculation that, despite the electron–electron and EP interactions each separately favouring insulating phases, together they can mediate an unexpected intermediate metallic state [5, 9]. Applying a variable-displacement Lang–Firsov scheme, a rather intriguing behaviour in the local magnetic moment and the renormalised hopping integral was found in the CDW-SDW cross-over region and traced back to the appearance of a metallic phase in the weak-coupling regime [5]. The intermediate phase has been confirmed by a numerical study based on stochastic series expansion quantum Monte Carlo with directed loops [9], where the slope of the finite-size scaled charge and spin structure factors in the long-wavelength limit was evaluated:

$$K_{\rho/\sigma} = \lim_{q\to 0} \frac{1}{\pi q} \frac{1}{N} \sum_{j,k} \mathrm{e}^{\mathrm{i}q(j-k)} \langle (n_{j,\uparrow} \pm n_{j,\downarrow})(n_{k,\uparrow} \pm n_{k,\downarrow}) \rangle .$$

$K_{\rho/\sigma}$ are the LL charge/spin correlation exponents. K_ρ values greater than 1 were taken as indication for dominant attractive superconducting correlations (SC). The resulting ground-state phase diagram shows two different sequences of phases as λ increases, either Mott–Peierls in the strong-coupling region of large u (as in discussed in Sect. 3) or Mott–SC–Peierls in the weak-coupling region of small u (see Fig. 1.11, upper column). At smaller phonon frequencies ω_0, the SC region shrinks. As it is numerically difficult to determine the phase boundaries by exploiting the local magnetic moment, the effective hopping integral or $K_{\rho/\sigma}$, we re-investigated the weak-coupling Holstein–Hubbard model by calculating the charge and spin gaps, defined in (1.14) and (1.15), respectively, as well as the two-particle excitation gap [42]

$$\Delta_{\mathrm{c}_2} = E_0^{2+}(0) + E_0^{2-}(0) - 2E_0(0). \tag{1.16}$$

This gap corresponds to the charge gap in a bipolaronic insulator and is an upper limit for it in any other phase. Let us denote the (single-particle) charge gap (1.14) by Δ_{c_1} in this section. Then, of course, one- and two-particle excitation gaps should simultaneously open if we enter the PI and MI phases. If the PI phase is a bipolaronic insulator (superlattice) rather than a traditional Peierls band insulator, mobile bipolarons may occur first in the dissolving process of the PI, as the λ/u ratio is lowered. Such a bipolaronic metal/liquid phase will then be characterised by $\Delta_{\mathrm{c}_2} = 0$ but finite Δ_{c_1} (and Δ_{s}). Adding/removing a single particle from the metallic bipolaron phase is energetically costly because the bipolarons are (tightly) bound. A bipolaron as a whole, however, can be added or removed without effort.

The left lower panel of Fig. 1.11 demonstrates that this scenario holds in the anti-adiabatic ($\omega_0/t = 5$) weak-coupling regime. As the EP coupling gets weaker, we enter a region where $\Delta_{\mathrm{c}_1} > 0$ but $\Delta_{\mathrm{c}_2} \to 0$ (see, e.g. the data for $\lambda = 0.625$). Δ_{c_2} stays zero as Δ_{c_1} vanishes at still smaller λ (u fixed),

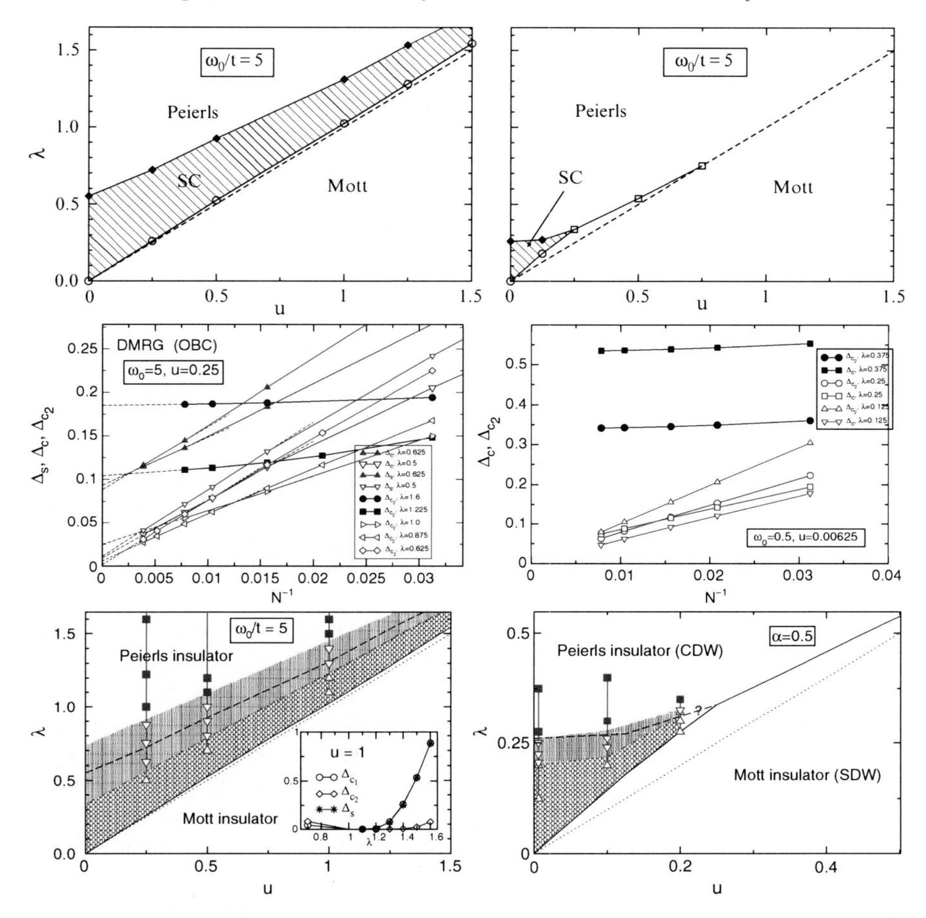

Fig. 1.11. Phase diagram of the Holstein–Hubbard model for weak couplings. The *left* (*right*) panels belong to the anti-adiabatic (adiabatic) regime. The *upper panels* are redrawn from Fig. 4 of [9], where the *shaded areas* show the extension of the intermediate metallic phase as determined from the behaviour of $K_{\rho/\sigma}$. The middle column panels display the finite-size scaling of spin, single-particle and two-particle excitation gaps. The lower column shows the phase diagram obtained from our DMRG calculation [42]. Here *filled squares*, open triangles down, and open triangles up, denote the PI, bipolaronic metal and Luttinger-liquid metal phases, respectively. The *inset* in the lower left panel gives the ($N \to \infty$) extrapolated values of the one-particle, two-particle and spin excitation gaps at $u = 1$. Note that Δ_{c_2} is twice as large as Δ_{c_1} in the MI phase. For further explanation see text

until we enter the MI state. We note that the PI-metal phase boundary is shifted compared to the results of [9], while the metal-MI transition line is the very same. That is, taking $\Delta_{c_2} = 0$ as a criterion for the instability of the PI phase, we find an even larger region for the PI-MI intervening state. Second, within the metallic state, our data indeed suggests a cross-over between a bipolaronic liquid ($\Delta_{c_2} = 0$; $\Delta_{c_1}, \Delta_s > 0$) and a Luttinger liquid ($\Delta_{c_2} = \Delta_{c_1} = \Delta_s = 0$). The corresponding results for the adiabatic regime ($\omega_0/t = 0.5$) are given in right lower panel of Fig. 1.11. Again we have strong evidence for an intermediate metallic state. The region where bound mobile charge carriers (bipolarons) exist, however, now is a small strip between the PI and metal phases only, and expected to vanish if the adiabaticity ratio ω_0/t goes to zero.

1.5 Limiting Cases

1.5.1 Adiabatic Holstein–Hubbard Model

In the adiabatic limit $\omega_0 = 0$, the general Holstein–Hubbard Hamiltonian (1.1) reduces to

$$H = H_{t-U} - \sum_{i,\sigma} \Delta_i n_{i\sigma} + \frac{\kappa}{2} \sum_i \Delta_i^2, \tag{1.17}$$

where H_{t-U} constitutes the conventional Hubbard Hamiltonian. In addition, the adiabatic Holstein–Hubbard model (1.17) includes the elastic energy of the lattice with 'spring constant' κ. Within this so-called 'frozen phonon' approach, $\Delta_i = (-1)^i \Delta$ is a measure of the static, staggered density modulations of the PI phase. For the adiabatic Holstein–Hubbard model a discontinuous PI–MI transition is easily verified in the atomic limit $t = 0$, where $\Delta = 1/\kappa$ for $U < U_c = 1/\kappa$ and $\Delta = 0$ for $U > U_c$. The first-order nature persists for finite small t, that is in the strong coupling regime U, $\kappa^{-1} \gg t$. However, we have demonstrated by an ED study that the transition is second order in the weak coupling regime U, $\kappa^{-1} \ll t$ [37]. This implies a continuous decrease of $\Delta(U)$.

We summarise our previous findings in the phase diagram shown in Fig. 1.12. In the Peierls BI phase for $U < U_{\mathrm{opt}}$, the spin and charge excitation gaps are equal and finite, and remarkably $\Delta_{\mathrm{opt}} \neq \Delta_c$. Here U_{opt} marks the point when the site-parity sectors become degenerate and the optical absorption gap Δ_{opt} disappears. At $U = U_{\mathrm{opt}}$, $\Delta_{\mathrm{opt}} = 0$ but $\Delta_c = \Delta_s > 0$. For $U \geq U_s$, the usual MI phase with $\Delta_{\mathrm{opt}} = \Delta_c > \Delta_s = 0$ is realised. For strong coupling $U_{\mathrm{opt}} = U_s$ holds. In weak coupling there exists an intermediate region $U_{\mathrm{opt}} < U < U_s$ in which all excitation gaps are finite. The CDW persists for all $U < U_s$. The site-parity eigenvalue is $P = +1$ in the PI and $P = -1$ in the MI phase. It is natural to expect an additional ordering phenomenon in the window $U_{\mathrm{opt}} < U < U_s$. Here a bond order wave (BOW) with a finite expectation value of the staggered bond charge $B = \frac{1}{N} \sum_{i\sigma} (-1)^i \langle c_{i\sigma}^\dagger c_{i+1\,\sigma} + \mathrm{H.c.} \rangle$ is the natural candidate.

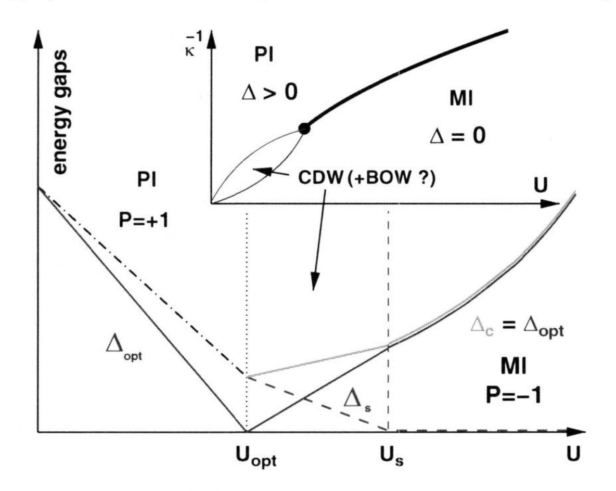

Fig. 1.12. Qualitative phase diagram of the adiabatic Holstein–Hubbard model. *Inset*: Transition scenario in the weak coupling regime $U, \kappa^{-1} \ll t$, where the PI-MI transition therefore evolves across two critical points

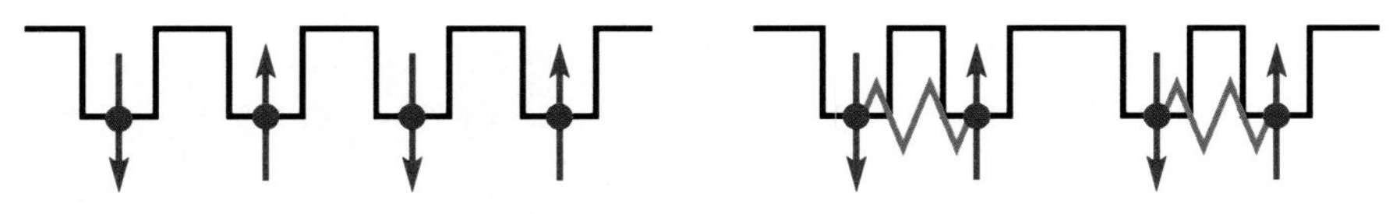

Fig. 1.13. Structure of the Mott SDW and spin–Peierls phases

1.5.2 Spin–Peierls Model

In the preceding sections we have seen that the interaction of electronic and vibrational degrees of freedom can lead to an instability of 1D metals towards lattice distortion. A similar effect is observed in quantum spin chains (which implement, in some sense, the $t = 0$ large U limit of the half-filled Hubbard model), where the coupling to the lattice can cause a so-called spin–Peierls transition from a spin liquid with gapless excitations to a dimerised phase with an excitation gap (see Fig. 1.13). Experimentally such behaviour was first observed in the 1970s for organic compounds of the TTF and TCNQ family [43]. The topic regained attention after the discovery of the first inorganic spin–Peierls compound $CuGeO_3$ in 1993 [44]. The most significant feature distinguishing this material from other spin–Peierls compounds is the high frequency ω_0 of the involved optical phonons, which is comparable to the magnetic exchange interaction J.

As an archetypal model for this type of spin–Peierls system we consider the anti-ferromagnetic Heisenberg chain coupled to optical phonons

$$H = J \sum_i \boldsymbol{S}_i \boldsymbol{S}_{i+1} + \omega_0 \sum_i b_i^\dagger b_i + H_{sp}, \tag{1.18}$$

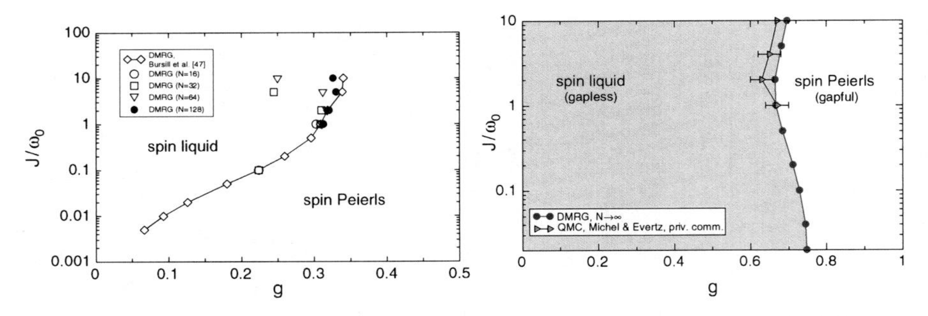

Fig. 1.14. Ground-state phase diagram of the spin–Peierls chain with difference (*left panel*) and local coupling (*right panel*)

where $\boldsymbol{S}_i$ denote spin-$\frac{1}{2}$ operators at lattice site i. For the spin–phonon interaction one usually considers two simple forms

$$H_{\text{sp}}^{d} = g\omega_0 \sum_i (b_i^{\dagger} + b_i)(\boldsymbol{S}_i\boldsymbol{S}_{i+1} - \boldsymbol{S}_{i-1}\boldsymbol{S}_i), \tag{1.19}$$

$$H_{\text{sp}}^{l} = g\omega_0 \sum_i (b_i^{\dagger} + b_i)\boldsymbol{S}_i\boldsymbol{S}_{i+1}. \tag{1.20}$$

H with the first (difference) type of spin–phonon interaction (1.19) has been studied with a large number of methods, including perturbation theory [45,46], flow equations [47], ED [46] and DMRG [48,49]. The latter approach identified the ground-state phase diagram displayed in Fig. 1.14 (left panel). All these studies agree on the main finding that at $\omega_0 > 0$ the system undergoes a QPT only for some finite value of g. The nature of the QPT from the gapless to the dimerised phase is rather well understood: For finite phonon frequency ω_0 the spin–phonon coupling g leads to effective spin interactions beyond nearest-neighbour exchange, that is the low energy physics is governed by a frustrated Heisenberg model, which has a gapped ground-state at sufficiently large frustration [50].

For the second (local) type of spin–phonon coupling (1.20), which applies to $CuGeO_3$, the precise location of the phase boundary has been determined only quite recently by a high-performance, parallel version of the well-known two-block finite-lattice DMRG algorithm [51] (see Fig. 1.14, right panel). To detect the quantum phase transition from the gapless to the dimerised phase, we used the established criterion of level-crossing between the first singlet and the first triplet excitation.

1.6 Conclusions

In summary, we have addressed the metal–insulator and insulator–insulator transition problem in 1D strongly coupled electron–phonon systems. The generic features observed are relevant to several classes of low-dimensional materials. Applying numerical diagonalisation methods based on Lanczos, density matrix renormalisation group and kernel polynomial algorithms, we analysed the general Holstein–Hubbard model at half-filling and obtained, by the use of present-day leading-edge supercomputers, basically exact results for both ground-state and spectral properties in the overall region of electron–electron/electron–phonon coupling strengths and phonon frequencies.

For the spinless Holstein model we found that for weak electron–phonon couplings the system resides in a metallic (gapless) phase described by two non-universal Luttinger-liquid parameters. Increasing the electron-phonon coupling, a quantum phase transition to a Peierls insulating state takes place, which is accompanied by drastic changes in the optical response of the system.

Of a similar type is a Heisenberg spin chain coupled to optical phonons, which for increasing spin-lattice coupling undergoes a quantum phase transition from a gapless spin liquid to a gapped phase with lattice dimerisation.

For the more involved Holstein–Hubbard model, with respect to the metal, the electron–electron interaction favours a Mott insulating state, while the electron–phonon coupling is responsible for the Peierls insulator to occur. True long-range (charge density wave) order is established in the Peierls insulator phase only. The Peierls insulator typifies a band insulator in the adiabatic weak-to-intermediate coupling range or a bipolaronic insulator for non-too-anti-adiabatic strong-coupling. The optical conductivity signals that the quantum phase transition between the Mott and Peierls insulator phases is connected to a change in the ground-state site-parity eigenvalue (of finite systems with PBC). While we found only one critical point separating Peierls and Mott insulating phases in the strong-coupling regime, there is strong evidence for an intervening metallic state in the weak-coupling regime. This is in accordance with results obtained in the adiabatic limit ($\omega_0 = 0$), where two successive transitions have been detected for weak couplings as well. The Peierls-to-Mott transition scenario is corroborated by the behaviour of the spin- and charge excitation gaps. From a DMRG finite-size scaling, we found that the charge gap equals the spin gap in the Peierls insulator while $\Delta_{\mathrm{c}} > \Delta_{\mathrm{s}} = 0$ in the Mott insulator, which proves spin–charge separation in the latter state.

Note added in proof. After this work was submitted the Tomonaga–Luttinger-liquid correlation parameter K_ρ has been determined for the half-filled spinless fermion Holstein model by DMRG, exploiting the static charge structure in the long–wavelength limit, $K_\rho = \pi \lim_{q \to 0^+} [S_c(q)/q]$, with $q = 2\pi/N$ ($N \to \infty$), rather than the leading–order scaling relations (1.4) and (1.5) [52]. While both approaches give almost identical results for intermediate–to–large phonon frequencies, the authors of Ref. [52] find $K_\rho < 1$ also in the adiabatic regime, which puts the subdivison of the metallic state into anattractive and repulsive Luttinger liquid into question.

Acknowledgements

We thank A. Alvermann, K.W. Becker, A.R. Bishop, F. Göhmann, M. Hohenadler, E. Jeckelmann, A.P. Kampf, A. Weiße and G. Wellein for valuable discussions. This work was supported by DFG through SFB 652 and KONWIHR Bavaria project HQS@HPC. Furthermore, we acknowledge generous computer granting by LRZ Munich and HLRN Berlin.

References

1. N.F. Mott, *Metal–Insulator Transitions* (Taylor and Francis, London, 1990)
2. R. Peierls, *Quantum theory of solids* (Oxford University Press, Oxford, 1955)
3. N. Tsuda, K. Nasu, A. Yanese, K. Siratori, *Electronic Conduction in Oxides* (Springer, Berlin, 1990)
4. A.R. Bishop, B.I. Swanson, Los Alamos Sci. **21**, 133 (1993)
5. Y. Takada, A. Chatterjee, Phys. Rev. Lett. **67**, 081102(R) (2003)
6. H. Fehske, G. Wellein, G. Hager, A. Weiße, A.R. Bishop, Phys. Rev. B **69**, 165115 (2004)
7. M. Capone, G. Sangiovanni, C. Castellani, C. Di Castro, M. Grilli, Phys. Rev. Lett. **92**, 106401 (2004)
8. G.S. Jeon, T.H. Park, J.H. Han, H.C. Lee, H.Y. Choi, Phys. Rev. Lett. **70**, 125114 (2004)
9. R.T. Clay, R.P. Hardikar, Phys. Rev. Lett. **95**, 096401 (2005)
10. M. Tezuka, R. Arita, H. Aoki, Phys. Rev. Lett. **95**, 226401 (2005)
11. T. Holstein, Ann. Phys. (N.Y.) **8**, 325 (1959)
12. J. Hubbard, Proc. R. Soc. Lond. Ser. A **277**, 237 (1964)
13. J.K. Cullum, R.A. Willoughby, *Lanczos Algorithms for Large Symmetric Eigenvalue Computations*, vol. I and II (Birkhäuser, Boston, 1985)
14. A. Weiße, G. Wellein, A. Alvermann, H. Fehske, Rev. Mod. Phys. **78**, 275 (2006)
15. S.R. White, Phys. Rev. Lett. **69**, 2863 (1992)
16. E. Jeckelmann, H. Fehske, Riv. Nuovo Cimento **30**, 259 (2007)
17. H. Fehske, A. Weiße, R. Schneider, (Eds.), *Computational Many-Particle Physics*, vol. 739 (Springer, Berlin Heidelberg, 2008)
18. J.E. Hirsch, E. Fradkin, Phys. Rev. B **27**, 4302 (1983)
19. G. Benfatto, G. Gallavotti, J.L. Lebowitz, Helv. Phys. Acta **68**, 312 (1995)
20. H. Zheng, D. Feinberg, M. Avignon, Phys. Rev. B **39**, 9405 (1989)
21. H. Fehske, M. Holicki, A. Weiße, in *Advances in Solid State Physics 40*, ed. by B. Kramer (Vieweg, Wiesbaden, 2000), pp. 235–249
22. R.H. McKenzie, C.J. Hamer, D.W. Murray, Phys. Rev. B **53**, 9676 (1996)
23. R.J. Bursill, R.H. McKenzie, C.J. Hamer, Phys. Rev. Lett. **80**, 5607 (1998)
24. G. Wellein, H. Fehske, Phys. Rev. B **58**, 6208 (1998)
25. S. Sykora, A. Hübsch, K.W. Becker, G. Wellein, H. Fehske, Phys. Rev. B **71**, 045112 (2005)
26. M. Hohenadler, G. Wellein, A.R. Bishop, A. Alvermann, H. Fehske, Phys. Rev. B **73**, 245120 (2006)
27. F.D.M. Haldane, Phys. Rev. Lett. **45**, 1358 (1980)
28. S. Tomonaga, Prog. Phys. Ther. **5**, 544 (1950)

29. J.M. Luttinger, J. Math. Phys. **4**, 1154 (1963)
30. J.L. Cardy, J. Phys. A **17**, L385 (1984)
31. H. Fehske, G. Wellein, G. Hager, A. Weiße, K.W. Becker, A.R. Bishop, Phys. B **359-361**, 699 (2005)
32. E. Jeckelmann, F. Gebhard, F.H.L. Essler, Phys. Rev. Lett. **85**, 3910 (2000)
33. E. Jeckelmann, Phys. Rev. Lett. **89**, 236401 (2002)
34. D. Sénéchal, D. Perez, D. Plouffe, Phys. Rev. B **66**, 075129 (2002)
35. C.E. Greffield, G. Sangiovanni, M. Capone, Eur. Phys. J. B **44**, 175 (2005)
36. E. Jeckelmann, C. Zhang, S.R. White, Phys. Rev. B **60**, 7950 (1999)
37. H. Fehske, A.P. Kampf, M. Sekania, G. Wellein, Eur. Phys. J. B **31**, 11 (2003)
38. W. Kohn, Phys. Rev. **133**, A171 (1964)
39. A.A. Ovchinnikov, Sov. Phys. JETP **30**, 1160 (1970)
40. H. Fehske, G. Wellein, A. Weiße, F. Göhmann, H. Büttner, A.R. Bishop, Phys. B **312–313**, 562 (2002)
41. N. Gidopoulos, S. Sorella, E. Tosatti, Eur. Phys. J. B **14**, 217 (2000)
42. H. Fehske, G. Hager, E. Jeckelmann, Europhys. Lett. **84**, 57001 (2008)
43. J.W. Bray, H.R. Hart Jr., L.V. Interrante, I.S. Jacobs, J.S. Kasper, G.D. Watkins, S.H. Wee, J.C. Bonner, Phys. Rev. Lett. **35**, 744 (1975)
44. M. Hase, I.Terasaki, K. Uchinokura, Phys. Rev. Lett. **70**, 3651 (1993)
45. K. Kuboki, H. Fukuyama, J. Phys. Soc. Jpn. **56**, 3126 (1987)
46. A. Weiße, G. Wellein, H. Fehske, Phys. Rev. B **60**, 6566 (1999)
47. C. Raas, A. Bühler, G.S. Uhrig, Eur. Phys. J. B **21**, 369 (2001)
48. R.J. Bursill, R.H. McKenzie, C.J. Hamer, Phys. Rev. Lett. **83**, 408 (1999)
49. G. Hager, A. Weiße, G. Wellein, E. Jeckelmann, H. Fehske, J. Magn. Magn. Mater. **310**, 1380 (2007)
50. K. Okamoto, K. Nomura, Phys. Lett. A **169**, 433 (1992)
51. A. Weiße, G. Hager, A.R. Bishop, H. Fehske, Phys. Rev. B **74**, 214426 (2006)
52. E. Ejima, H. Fehske, Europhys. Lett. **87**, 27001 (2009)

2

The Metal–Nonmetal Transition in Fluid Mercury: Landau–Zeldovich Revisited

Friedrich Hensel

Abstract. The paper reviews recent experimental results, which show that the formation of a mixed phase of metallic and nonmetallic domains following the putative 'Landau–Zeldovich first-order metal–nonmetal transition' noticeably influences the liquid–vapor transition of fluid mercury. An attempt is made to connect the observable consequences of the mixed phase existence with the scenario of noncongruent evaporation.

2.1 Introduction

An important open problem in the statistical mechanics of fluids is an understanding of the interrelation of the metal–nonmetal transition and the liquid–vapor phase transition in metallic systems. It is now 50 years since Landau and Zeldovich [1] first called attention to this question, specifically in relation to liquid mercury. While emphasizing that one cannot distinguish metal (M) from nonmetal (NM) above the absolute zero of temperature, they nevertheless proposed that electronic transitions could introduce additional lines of first-order, that is, discontinuous electronic transitions in the phase diagram of the fluid state. They suggested the three possible diagrams shown in Fig. 2.1. Lines of first-order metal–nonmetal transitions might occur wholly in the liquid (Fig. 2.1b) or vapor state (Fig. 2.1c) or, as a third possibility, the transition might be an extension of the liquid–vapor pressure curve (Fig. 2.1a) beyond the critical point into the supercritical state. Specifically for mercury, Landau and Zeldovich [1] propose Fig. 2.1b with a triple point at the intersection of the liquid–vapor and the metal–nonmetal transition.

The problem that throughout the fluid range of mercury the electronic structure can change discontinuously with temperature and pressure has stimulated extensive experimental and theoretical efforts and the general subject has been repeatedly reviewed in the literature [2–10]. A comprehensive review of the available experimental and theoretical results is therefore unnecessary, and I shall refrain from any attempt to cover the entire field. Instead, I have

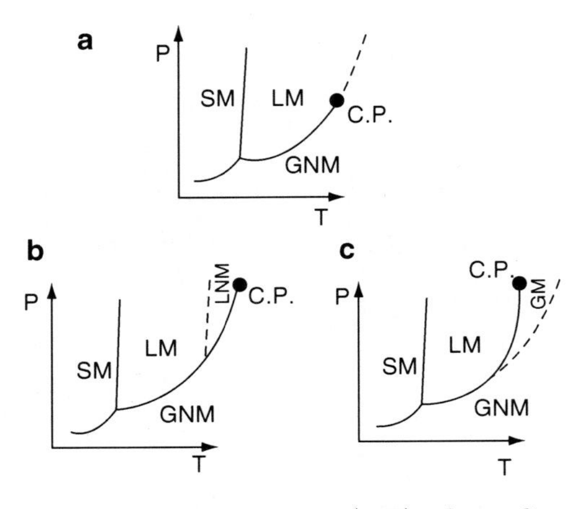

Fig. 2.1. Schematic pressure–temperature (p-T) phase diagrams proposed by Landau and Zeldovich [1]: S = solid, L = liquid, G = gas, M = metal, NM = non-metal. The critical point C.P. terminates the line of liquid–gas equilibrium. *Dashed curves* indicate lines of proposed first-order M–NM transitions

selected for attention recent, partly unpublished, experimental results which evidently show that the apparently continuous changes in the electric structure and transport properties of fluid mercury observed during the course of the metal–nonmetal (M–NM) transition are connected with local microscopic inhomogeneities in the electronic structure.

The renewed interest in this problem is mainly motivated by the exciting progress in the search for density fluctuations inherent to the M–NM transition in expanded fluid mercury, which has come from the recent small angle X-ray scattering (SAXS) measurements of Inui et al. [11–13] and from very accurate ultrasound velocity [14] and absorption [15, 16] measurements by Yao and coworkers and by Kozhevnikov [17].

2.2 The Liquid–Vapor Phase Boundary of Mercury

The SAXS experiments on mercury under extreme conditions of temperature and pressure in the vicinity of the critical point and the M–NM transition were carried out on the high-energy X-ray diffraction beam line (BL04B2) at SPring-8 in Japan [13]. Inui et al., were able to observe a change in the character of the SAXS-intensity as the density of liquid mercury increases from the critical region to higher densities where a continuous change from nonmetallic to metallic behavior has been predicted. Analysis of the data employing the usual Ornstein–Zernike plots shows that the long wavelength limit of the scattering function $S(0)$ and the correlation length ξ follow approximately through the peak around the critical point, but with increasing density $S(0)$

continues to decrease while ξ falls to a constant value, independent of temperature, of about 5–6 Å, indicating fluctuations of different character, which Inui et al. [13] ascribed correctly to fluctuations between insulating and metallic regions on a length scale of intermediate-range order. However, the Ornstein–Zernike analysis does not permit to truly separate the additional scattering related to the M–NM transition from the critical scattering. Be that as it may, the exciting observation of Inui et al. [13] stimulated Maruyama et al. [18] to apply the reverse Monte Carlo [19] method and the Voronoi–Delaunay analysis [20] on the basis of the structure factors determined by X-ray diffraction experiments by Tamura and Hosokawa [21] to characterize the intermediate-range fluctuations in the M–NM-range. The model structure resulting from the void analysis shows a well defined binary mixture of M- and NM-domains with arbitrary sizes and shapes. The concentration of the M-domains increases with increasing density and the spatial distribution of these domains resembles micro- or better nanoemulsions, that is the system can be treated as a two-density model. The assumption is then, that there is a matrix of density ρ_1 in which are embedded particles or domains of density ρ_2. More generally, the density in the system is either ρ_1 or ρ_2. If the two phases have sharp boundaries, there are general rules called the Porod law and the Porod invariant [22]. And as a matter of fact, a new inspection of the SAXS-data [23] over the density range from $3.5\,\mathrm{g\,cm^{-3}}$ to $11\,\mathrm{g\,cm^{-3}}$ covered by the experiment [11,12] shows that the data are consistent with Porods theory of two-phase systems [22]. In particular, in the liquid mercury density range between about 11 and $9\,\mathrm{g\,cm^{-3}}$, the coexistence of M nanodomains with a density of $10.7\,\mathrm{g\,cm^{-3}}$ and NM-nanodomains with a density of $8.3\,\mathrm{g\,cm^{-3}}$ is obtained.

As pointed out by Landau and Zeldovich [1], a salient feature concerning the M–NM transition in mercury is the possible occurrence of a first order phase transition, characterized by a discontinuous change in density from ρ_{M} to ρ_{NM}, that is there exists an interval in which the equilibrium state is macroscopically phase-separated into regions of higher and lower than average density. However, when long range Coulomb forces are taken into account, this instability with macroscopic separation is frustrated due to electrostatic energy cost. A Coulomb interaction precludes macroscopic phase separation: consequently, the system can form intermediate phases 'electronic micro- or nanoemulsions,' where domains of one phase (M) are embedded into the other phase (NM). A large number of small domains would minimize the Coulomb energy, but they cost too much surface energy. The distance between the domains and their size are determined by minimizing a free energy, which takes into account both the effects [24,25].

It goes without saying that the finding of the formation of a nanoemulsion in fluid mercury for densities smaller than about $11\,\mathrm{g\,cm^{-3}}$ has implications for the interpretation of the phase behavior and the electronic transport properties. In particular, liquid mercury can no longer be considered as a homogeneous one-component fluid for which the liquid–vapor phase boundary in the pressure (p)–temperature (T) plane is represented by a single line, the vapor pressure curve. If it forms a binary mixture of M- and NM domains, the general features of its liquid–vapor transition are known from the thermodynamics of mixtures; that is, the two-phase region in the p–T-plane is no longer a single line but a two-dimensional domain, whose boundary parameters depend on the [NM]/[M] concentration ratio. Before turning to the discussion of new results, it is inevitable that some steps of the previous reviews will be retraced in order that the present account should not be unsystematic.

There used to be general agreement that the most significant experiments relevant to the exploration of the relationship between the liquid–vapor and metal–nonmetal transitions in fluid mercury are direct measurements of electrical properties that signal the transformation from a metallic to a nonmetallic state. Data such as those of the electrical conductivity (see Fig. 2.2) [26–30] clearly demonstrated that for fluid mercury there is no sharp (first-order) electronic transition except across the apparent liquid–vapor phase boundary, that is, the liquid–vapor phase separation tends to separate the nonmetallic and metallic fluids. Near the apparent critical point the conductivity drops sharply, thus showing a strong effect of the incipient phase transition on the electronic structure. The close correlation between the behavior of the density and that of the electrical conductivity (Fig. 2.2) convincingly shows that the variation of elemental density is the dominant factor governing the metal–nonmetal transition. However, it has to be emphasized that in practice very careful measurements are required to separate the ever present effects of density and temperature in the apparent critical region. Part of the difficulty arises because both the compressibility and also the pressure derivative of the electrical conductivity become very large in the critical region. This means that small errors in pressure measurement will cause large density and conductivity errors. Consequently, for a reliable correlation of the conductivity and density, precise temperature and pressure control is essential. This is not easily achievable, because precise measurements of the properties of fluid metals are notoriously difficult. This is immediately evident from the fact that the critical point of mercury is near a temperature $T = 1{,}478°\mathrm{C}$ and a pressure $p = 1{,}673$ bar well beyond the range of standard experimental techniques of condensed matter physics.

A serious additional experimental problem related to the determination of the liquid–vapor phase boundary in mercury is that nearly all investigations – including the experimental data in Fig. 2.2 – are not concerned with saturation conditions where both liquid and vapor are present in equilibrium. Instead, the method usually employed was to heat a sample continuously at a constant pressure until an abrupt change of the measured property (e.g., the conductivity) signals that apparently the liquid boiled out of the cell at the vaporization point. This is a very efficient method and the vapor pressure data obtained in this way are very accurate and reproducible. However, it goes without saying that the employment of this method by nearly all experimentalists working on the determination of the liquid–vapor phase boundary of mercury is based on the assumption that mercury forms a homogeneous one-component

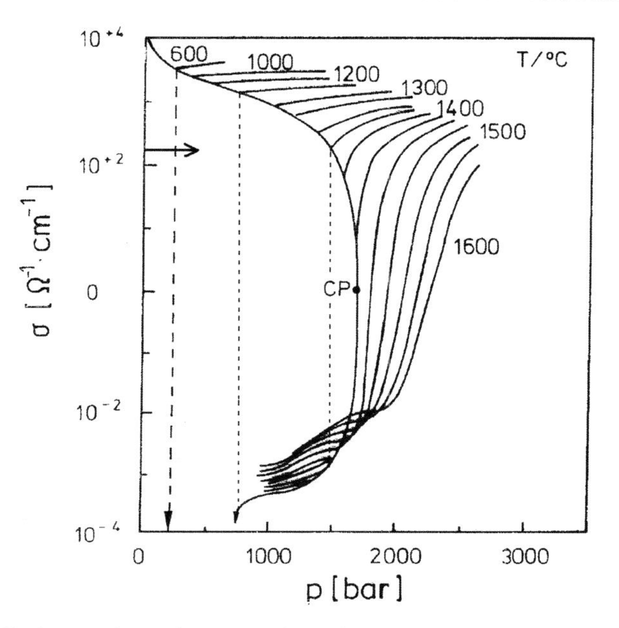

Fig. 2.2. DC electrical conductivity data for mercury at sub- and supercritical temperatures as a function of pressure [26–29]

liquid; thus, ignoring the fact that the thermal equilibrium times in a mixture liquid–vapor two-phase region are much longer than those of pure substances for a given temperature distance ΔT from the transition [31]. There is only one experiment described in the literature that was mainly concerned with saturation conditions where both liquid and vapor were present in equilibrium [32, 33]. The authors of this work – also assuming that mercury is a homogeneous one-component liquid over the whole liquid range – pointed out that their vapor pressure curve $p(T)$ of mercury – measured under true equilibrium conditions – has an unusual and probably unique form in that the logarithm of the pressure (p) against the reciprocal temperature ($1/T$) plot showed a relatively sharp change of slope at a temperature of 1,088°C (1,361 K) and a saturated liquid density $\rho_{\mathrm{L}} = 10.7\,\mathrm{g\,cm^{-3}}$. As mentioned earlier, the latter is about the density at which the inspection [23] of the SAXS- data [11] shows that liquid mercury starts to form a nanoemulsion.

Seen at glance from Fig. 2.3, the two experimentally determined curves BC and SC (solid lines) do not intersect. The dashed line (see *inset*) is only an attempt (assuming that the two curves approach each other) to compare the phase behavior of mercury with well-known features from the theory of fluid binary mixtures [34], that is cricondentherm $T_{\max}$, cricondenbar $p_{\max}$, and the critical point C.P., for which the composition of the two coexisting phases is equal. The total vapor pressure within the two-phase region of mercury is plotted as a function of density in form of isobars or isotherms, respectively, in

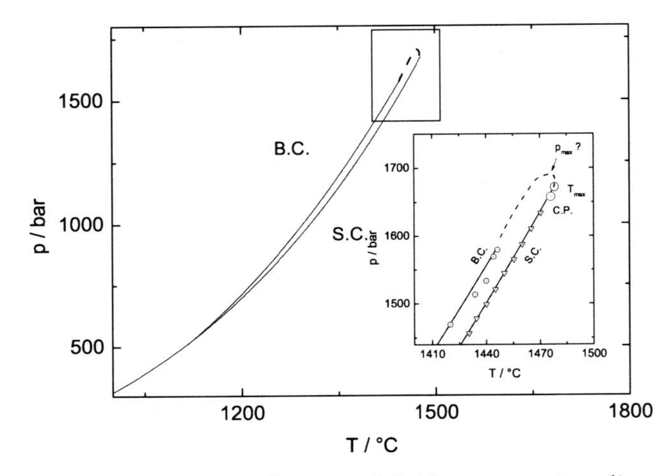

Fig. 2.3. Pressure–temperature diagram of fluid mercury: two-dimensional two-phase region for temperatures higher than $T = 1,088°$C is observed instead of the standard p–T saturation curve; *BC*-boiling (liquid) conditions, *C.P.* critical point; $T_{\max}$ = cricondentherm, $p_{\max}$ = cricondenbar. As the critical point of binary systems with variable composition is difficult to locate, we choose the location at $([\mathrm{NM}]/[\mathrm{M}])_{\mathrm{BC}} = ([\mathrm{NM}]/[\mathrm{M}]_{\mathrm{SC}})$

Figs. 2.4 and 2.5. The isothermic phase transition starts and finishes at different pressures, while the isobaric phase transition starts and finishes at different temperatures. The curves inside the two phase regions are not really straight lines, and they should be simply considered as a guide to the eyes. They are horizontal only for temperatures smaller than 1,088 °C or pressures smaller than 458 bar, respectively.

In Fig. 2.3, we displayed both vapor pressure branches: the pressure obtained for slow evaporation [32,33] under true equilibrium conditions resulting in the bubble or boiling curve, which we designate BC together with the pressure obtained for nonequilibrium conditions when temperature or pressure are changed with a finite rate [26]. The latter results in the saturation curve, which we designate SC. The unusual form of the phase boundary has implications for the behavior of nearly all properties of mercury, which have not been recognized in the past. The essential difficulty is that the effect is small. This is immediately evident from the effect of isothermal evaporation on the DC electrical conductivity σ, which has been measured at BC condition [32] and SC-condition [26]. Figure 2.6 displays the effects of evaporation for the two isotherms $T = 1{,}350°$C, and $T = 1{,}400°$C. The differences in pressure for SC- and BC-conditions are not large enough to be seen in a diagram as that shown in Fig. 2.2. As the electrical conductivity under these conditions can be well described [36], by the effective medium approach [35] we are able to calculate the volume fraction ϕ of the metallic component at liquid BC- and SC conditions. The values of ϕ are displayed in the plot of Fig. 2.6. Similar effects are observed for evaporation at the isobar of 1,400 bar (see Fig. 2.7).

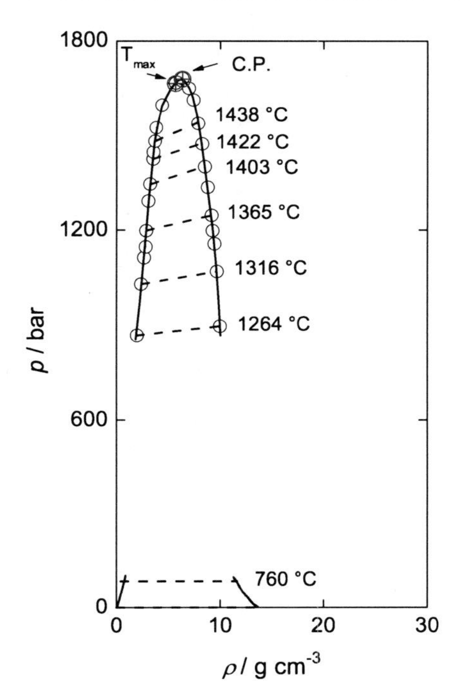

Fig. 2.4. Non-congruent pressure density diagram. For temperatures higher than 1,088°C, the isothermal phase transition starts and finishes at different pressures. The dashed lines are not measured curves but serve only as a guide for the eyes. BC = boiling curve, SC = saturation curve; $T_{\max}$ = cricondentherm. C.P. critical point where $([\mathrm{NM}]/[\mathrm{M}])_{\mathrm{BC}} = ([\mathrm{NM}]/[\mathrm{M}])_{\mathrm{SC}}$

The availability of electrical conductivity data at BC- and SC-conditions permits us to calculate the volume fractions of the components by employing the effective medium theory. Data for the volume fraction of the metallic component are presented in Fig. 2.8 in the form of isobars as a function of temperature. The difference in the volume fraction at liquid BC- and liquid SC-condition is relatively large. The dashed line is based on the assumption that the volume fractions for SC- and BC-condition is equal at the 'apparent' critical point.

The scenario is thus very similar to that of noncongruent phase transitions highlighted by Iosilevskiy and colleagues [37–39] in relation to the evaporation of uranium dioxide. The effect of the noncongruent phase transition is most vividly seen in the shape of the liquid vapor coexisting curve. Figure 2.9 shows an extension of the density–temperature plot of Fig. 2.5 over the whole liquid–vapor coexisting range. The most remarkable feature is the strong deviation from the empirical rule of rectilinear diameters of Cailletet and Mathias [40].

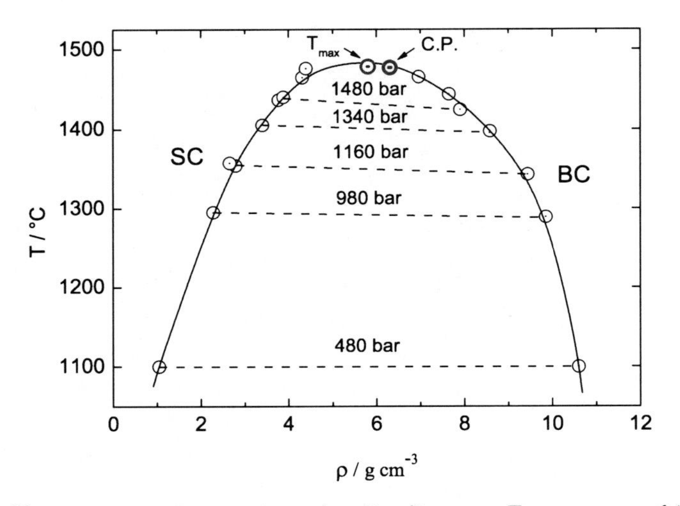

Fig. 2.5. Noncongruent temperature density diagram. For pressures higher than about 450 bar, the isobaric phase transition starts and finishes at different temperatures. The *dashed lines* are not measured curves but serve only as a guide for the eyes. *BC* = boiling curve, *SC* = saturation curve, $T_{\max}$ = cricondentherm

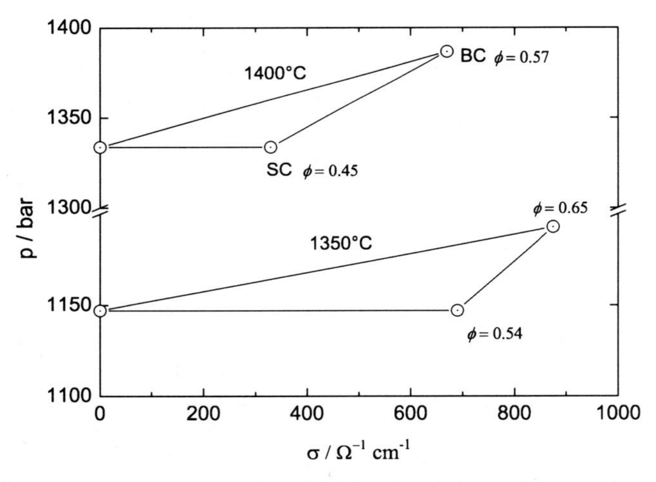

Fig. 2.6. Noncongruent pressure–electrical conductivity σ diagram for $T = 1{,}350°\mathrm{C}$ and $T = 1{,}400°\mathrm{C}$, ϕ is the volume fraction of the metallic component. The conductivity of the coexisting vapor is smaller than $10^{-3}\,\mathrm{ohm}^{-1}\,\mathrm{cm}^{-1}$

A remarkable wiggle is observed at about a temperature of 1,088°C. It is obvious that a law of corresponding states is not valid.

With the knowledge of the data in Figs. 2.4, 2.5 and 2.8, it is easy to calculate different properties that are effected by noncongruent evaporation. As an

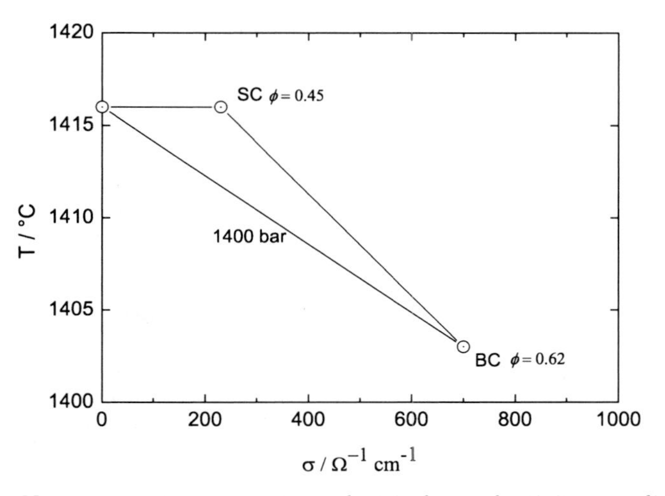

Fig. 2.7. Noncongruent temperature–electrical conductivity σ diagram for $p = 1{,}400$ bar, ϕ is the volume fraction of the metallic component. The conductivity of the coexisting vapor is smaller than $10^{-3}\,\mathrm{ohm}^{-1}\,\mathrm{cm}^{-1}$

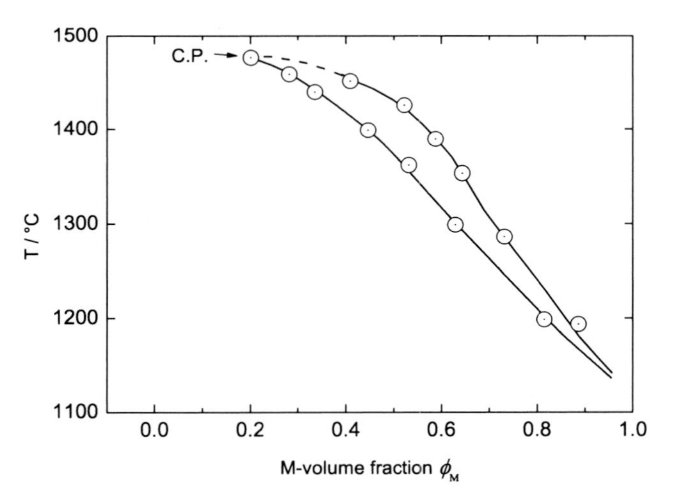

Fig. 2.8. Volume fraction ϕ of the metallic component at BC liquid and SC liquid conditions. ϕ is calculated from the measured electrical conductivity employing the effective medium theory [35]. The *dotted line* is an extrapolation assuming that ϕ along the BC- and SC- curves becomes equal at C.P

example, we calculate for the density of $9.45\,\mathrm{g\,cm^{-3}}$ at the liquid BC-branch different derivatives

$$\left(\frac{\partial \rho}{\partial T}\right)_{\mathrm{BC}} = 0.007\,\frac{\mathrm{g}}{\mathrm{cm^3K}};\ \left(\frac{\partial p}{\partial \phi}\right)_{\rho,T} = 669\,\mathrm{bar};\ \left(\frac{\partial \phi}{\partial \rho}\right)_{p,T} = -0.4\,\frac{\mathrm{cm^3}}{\mathrm{g}}$$

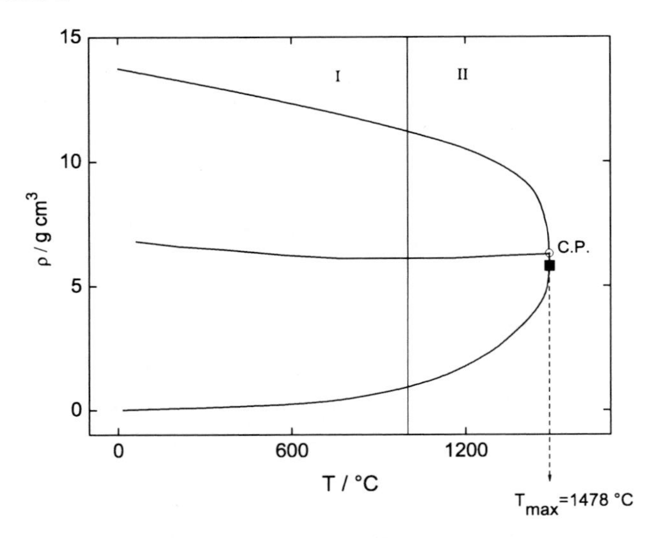

Fig. 2.9. Noncongruent evaporation of liquid mercury causes a pronounced wiggle of the diameter $\rho_\mathrm{d} = (\rho_\mathrm{l} + \rho_\mathrm{v})/2$ at a temperature of about 1,088°C with a skewing toward higher densities

from which we now evaluate the pressure derivative $\left(\frac{\partial p}{\partial T}\right)_{\mathrm{BC},\rho}$ of the $9.45\,\mathrm{g\,cm^{-3}}$ isochore close to BC-condition permitting variation of the composition ϕ as it occurs during bubble formation and the $\left(\frac{\partial p}{\partial T}\right)_{\mathrm{BC},\phi,\rho}$, keeping now the composition constant (the nonequilibrium fast measurement) from the thermodynamic relation:

$$\left(\frac{\partial p}{\partial T}\right)_{\mathrm{BC},\phi,\rho} - \left(\frac{\partial p}{\partial T}\right)_{\mathrm{BC},\rho} = \left(\frac{\partial \phi}{\partial \rho}\right)_{p,T} \left(\frac{\partial p}{\partial \phi}\right)_{\rho,T} \left(\frac{\partial \rho}{\partial T}\right)_{\mathrm{BC}}.$$

The calculated value for this difference is about $1.85\,\mathrm{bar\,K^{-1}}$.

This value is in very close agreement with the experimentally observed result [26, 32] displayed in Fig. 2.10, which is about $1.8\,\mathrm{bar\,K^{-1}}$. Another effect that most probably can be explained in terms of "noncongruent" evaporation is the observation of "anomalous" wiggles in the ultrasound velocity c as a function of pressure at constant temperature or alternatively as a function of temperature at constant pressure by Kobayashi et al. [14] at BC- and SC-condition. Figures 2.11 and 2.12 display the effect of isobaric and isothermal "noncongruent" evaporation for the experimentally determined adiabatic compressibility $\beta_\mathrm{S} = 1/\rho \cdot c^2$. The p–T coordinates observed by Kobayashi et al., for BC- and SC-conditions are included in the inset of Fig. 2.3 (the open circles and open triangles). They are surprisingly close to the vapor pressure curves found in [26, 32].

In conclusion, we regard the experimental observations presented above as compelling evidence that the features of the liquid mercury evaporation for

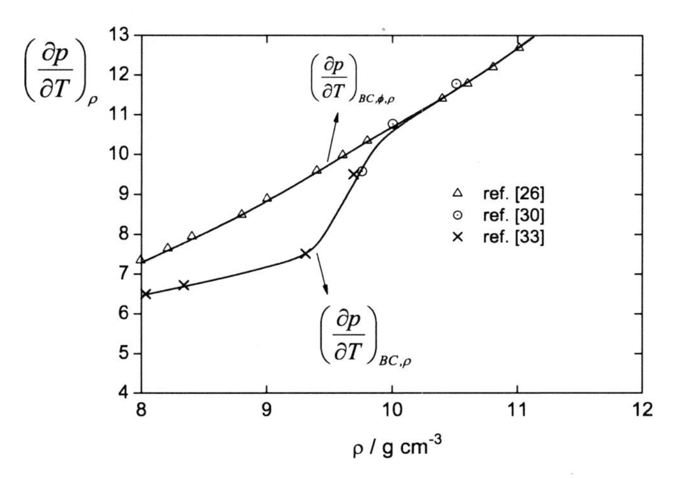

Fig. 2.10. Comparison between the measured isochoric pressure derivates $\left(\frac{\partial p}{\partial T}\right)_{\mathrm{BC},\phi,\rho}$ (*lower curve*) and $\left(\frac{\partial p}{\partial T}\right)_{\mathrm{BC},\rho}$ (*upper curve*) asymptotically close to BC. The two curves split into two branches for densities lower than $10.7\,\mathrm{g\,cm^{-3}}$ (see text)

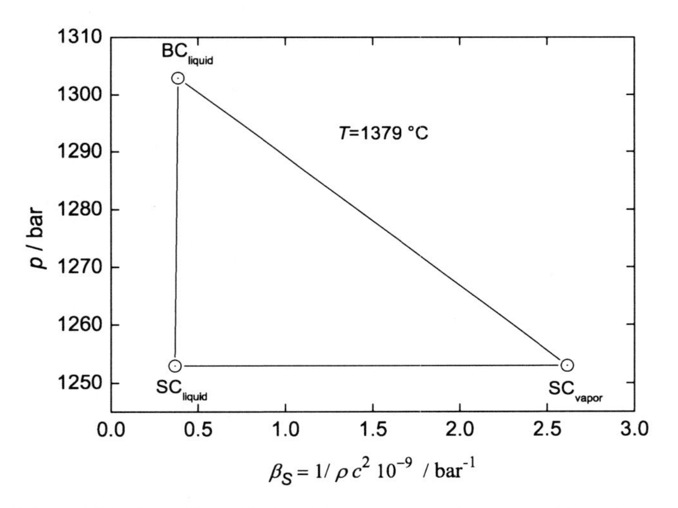

Fig. 2.11. Figure displays the effect of noncongruent evaporation at constant temperature on the adiabatic compressibility $\beta_\mathrm{S} = 1/\rho \cdot c^2$ at BC- and SC-conditions

temperatures higher than 1,088°C resemble the "noncongruent-evaporation" discussed by Iosilevskiy and colleagues [37–39]. This phenomenon may play an important role for many scenarios of postulated first-order phase transitions as those of the liquid–vapor transition in the high temperature uranium-oxide system [37], the postulated hypothetical plasma phase transition in giant planets [41], the phase separation in complex dusty plasmas which are

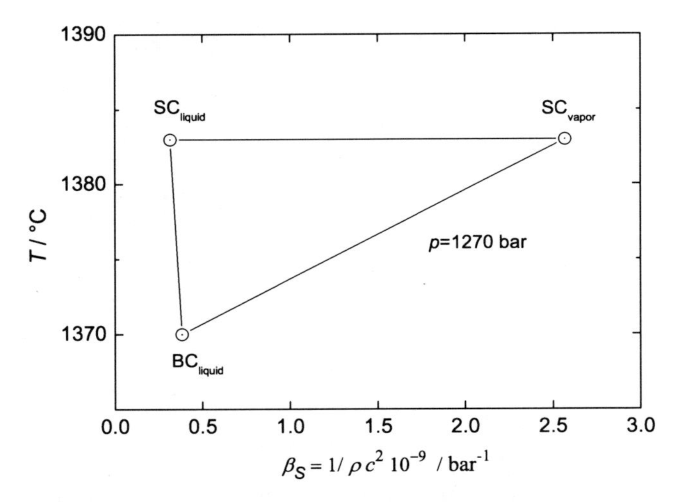

Fig. 2.12. Figure displays the effect of "noncongruent" evaporation at constant pressure on the adiabatic compressibility $\beta_S = 1/\rho \cdot c^2$ at BC- and SC-conditions.

predicted to show gas, fluid, and mixed phases of dust grains [42, 43], the formation of "pasta" structures in compact stars following an exotic first-order-phase transition [44], and the exotic phase transitions in neutron stars [45–47]. To our knowledge, the postulated noncongruence of hypothetical transitions in dense ionized systems at extreme conditions of temperature and pressure has never before been seen experimentally. If mercury is the first example for which the long-sought-for effect has been experimentally studied, it may serve as a model systems for simulating the transitions mentioned above for which the outlook for reliable measurements is unfavorable.

References

1. L. Landau, J. Zeldovich, Acta PhysicoChim. USSR **18**, 194 (1943)
2. F. Hensel, W.W. Warren Jr., *Fluid Metals* (Princeton University Press, Princeton, NJ, 1999)
3. H. Endo, Progr. Theor. Phys. Suppl. **72**, 100 (1982)
4. N.E. Cusack, in *The Metal Non-Metal Transition in Disordered Systems*, ed. by L.R. Friedmann, D.P. Tunstall, (Scottish University, Edinburgh, 1978) pp. 3
5. F. Hensel, H. Uchtmann, Ann. Rev. Phys. Chem. **40**, 61 (1989)
6. V.A. Alekseev, I.T. Iakubov, Phys. Rep. **96**, 1 (1983)
7. F. Hensel, Adv. Phys. **44**, 3 (1995)
8. F. Hensel, Phil. Trans. Roy. Soc. Lond. **356**, 97 (1998)
9. P.P. Edwards, R.L. Johnston, F. Hensel, C.N.R. Rao, P.P. Tunstall, Solid State Phys. **52**, 229 (1999)
10. F. Yonezawa, T. Ogawa, Prog. Theo. Phys. Suppl. **72**, 1 (1982)
11. M. Inui, K. Matsuda, D. Ishikawa, K. Tamura, Y. Ohishi, Phys. Rev. Lett. **98**, 185504 (2007)

12. M. Inui, K. Matsuda, K. Tamura, D. Ishikawa, J. Cryst. Soc. Jpn. **48**, 76 (2006)
13. M. Inui, K. Matsuda, K. Tamura, *SPring-8 Research Frontiers*, (Springer, Japan, 2007) p. 68
14. K. Kobayashi, H. Koyikawa, Y. Hiejima, T. Hoshino, M. Yao, J. Non-cryst. Solids **353**, 3362 (2007)
15. H. Kohno, M. Yao, J. Phys. Condens. Matter **13**, 10293 (2001)
16. H. Kohno, M. Yao, J. Phys. Condens. Matter **11**, 5399 (1999)
17. V. Kozhevniokov, Fluid Phase Equilib. **185**, 315 (2001)
18. K. Mayurama, H. Endo, H. Hoshino, F. Hensel, Phys. Rev. B. **80**, 014201-1 (2009)
19. R.L. McGreevy, J.Phys. Condens. Matter **13**, R877 (2001)
20. S.L. Chan, S.R. Elliott, Phys. Rev. B **43**, 4423 (1991)
21. K. Tamura, S. Hosokawa, Phys. Rev. B **58**, 9030 (1998)
22. G. Porod, Kolloid-Z. **124**, 83 (1951)
23. W. Ruland, F. Hensel, J. of Appl. Cryst. (2009) (accepted)
24. E. Nagaev, *Physics of Magnetic Semiconductors* (MIR, Moscow, 1983)
25. E. Nagaev, A.I. Podelschikov, V.E. Zilbewarp, J.Phys. Condens. Matter **10**, 9823 (1998)
26. W. Götzlaff, Dissertation, University of Marburg, 1988
27. F. Hensel, E.U. Franck, Ber. Bunsenges. Phys. Chem. **70**, 1154 (1966)
28. I.K. Kikoin, A.R. Sechenkov, Phys. Met. Metallogr. **24**, 74 (1967)
29. F.E. Neale, N.E. Cusack, J. Phys. E Sci. Instrum. **10**, 609 (1977)
30. M. Yao, H. Endo, J. Phys. Soc. Jpn. **51**, 966 (1982)
31. G.R. Brown, H. Meyer, Phys. Rev. A **6**, 1578 (1972)
32. S.R. Hubbard, R.G. Ross, J. Phys. C Solid State Phys. **16**, 6921 (1983)
33. S.R. Hubbard, R.G. Ross, Nature **295**, 682 (1973)
34. J.S. Rowlinson, F.L. Swinton, *Liquids and Liquid Mixtures* (Butterworth, London, 1982)
35. M.H. Cohen, J. Jortner, Phys. Rev. Lett. **30**, 695 (1973)
36. F. Hensel (to be published)
37. I. Iosilevskiy, E. Yakub, C. Ronchi. Int. J. Thermophys. **22**, 1253 (2001)
38. I. Iosilevskiy, V. Gryaznov, E. Yakub, C. Ronchi, V. Fortov, Contrib. Plasma Phys. **43**, 316 (2003)
39. I. Iosilevskiy, E. Yakub, G. Hyland, C. Ronchi, Trans. Am. Nucl. Soc. **81**, 122 (1999)
40. L. Cailletet, E.C. Mathias, Compt. Rendus Acad. Sci. **102**, 1202 (1886)
41. D. Saumon, G. Chabrier, W.B. Hubbard, J.I. Lunine, in *Strongly Coupled Plasma Physics*, ed. by M.H. van Horn, S. Ichimaru (University of Rochester Press, USA, 1993), p. 111
42. K. Avinash, Phys. Plasma. **8**, 2601 (2001)
43. G. Joyce, M. Lampe, G. Gauguli, Phys. Rev. Lett. **88**, 095006 (2002)
44. T. Marujasma, T. Tatsumi, T. Endo, S. Chiba, arXiv: nucl-th/060507v2 (2006)
45. H. Heiselberg, M. Hjorth-Jensen, arXiv: astro-ph/9802028v1 (1998)
46. N.K. Glendenning, Phys. Rep. **342**, 393 (2001)
47. N.K. Glendenning, Phys. Rev. C **47**, 2733 (1988)

3

The Influence of Pauli Blocking Effects on the Mott Transition in Dense Hydrogen

W. Ebeling, D. Blaschke, R. Redmer, H. Reinholz, and G. Röpke

Abstract. We investigate the effects of Pauli blocking on the properties of hydrogen at high pressures. In this region recent experiments have shown a transition from insulating behavior to metal-like conductivity. To describe this transition, several effects have to be taken into account, an important one is the quantum character of the electrons. As electron states can only be occupied once (Pauli blocking), atomic states need more phase space than available at high densities, and bound states disintegrate subsequently (Mott effect). We calculate the energy shifts due to Pauli blocking and discuss the Mott effect solving an effective Schrödinger equation for strongly correlated systems. Additionally, we include corrections due to polarization effects. The ionization equilibrium is treated on the basis of an advanced chemical approach based on the assumption that the system is a gas-like mixture of chemical species. We calculate the Pauli shifts by variational methods and discuss corrections due to polarization. Results for the ionization equilibrium in the region $5{,}000 < T[\mathrm{K}] < 15{,}000$, $0.1 < \rho[\mathrm{g\,cm^{-3}}] < 1$ are presented, where the transition from a neutral hydrogen gas to a highly ionized plasma occurs. We show that the transition to a highly conducting state is softer than predicted in earlier work.

3.1 Introduction

The physical properties of dense hydrogen is a topic that was raised in the past by many authors, starting with Wigner, Huntington, Abrikosov, and others [1,2]. Of special interest is the transition of hydrogen to a highly conducting phase, which is considered to be a type of Mott transition. It is well known and will also be illustrated in this volume that the Mott transition has characteristic features that are apparent for each relevant substance and also has a variety of specific aspects. In this contribution, we study the effects of Pauli's exclusion principle in dense hydrogen as the simplest representative of dense matter. Previous studies on dense hydrogen within advanced chemical models included several hypothetical assumptions about the character of the high-density phase [3–10]. Many questions remained open, in particular about the nature of the highly conducting, possibly even superconducting state.

Although metalization of solid hydrogen near $T = 0\,\mathrm{K}$ has not been clearly verified so far for pressures of up to 300 GPa [11], metallic-like features have been observed in shock-compression experiments using a two-stage light gas gun on fluid hydrogen and deuterium at Mbar pressures (1 Mbar = 100 GPa) and finite temperatures [11–13]. Metal-like conductivities have been observed around 140 GPa and 3,000 K [12]. Furthermore, significant discrepancies between the Hugoniot curves derived from laser-driven shock-wave experiments [11, 13] and theoretical equations of state such as the Sesame tables [3] have been found in the Mbar region where fluid hydrogen shows probably a larger compressibility than predicted. Other very recent experiments were able to reach that region as well and provided detailed information on the EOS and the conductivity in the Mbar region [14]. Both findings, the transition to metallic-like behavior and the increased compressibility, change our present understanding of the behavior of hydrogen at ultra-high pressures. It is relevant for models of planetary (see [15] and references therein) and stellar interiors [16] as well as for inertial confinement fusion studies [17].

In the present work, we show that one of the most important effects leading to the destruction of bound states like atoms and molecules is Pauli blocking. Because of the Pauli exclusion effect, the free electrons in the plasma cannot penetrate into the interior of atoms and molecules as the bound states are already occupied by the atomic electrons. At high densities this leads to an enormous pressure acting on the neutrals, which will finally be ionized. Several approaches to incorporate this effect will be discussed, starting from perturbation theory as well as finding variational solutions of effective wave equations, including Pauli blocking terms. Polarization effects will be included as well.

The effective energy levels of hydrogen, which depend on density and weakly also on temperature, are introduced into the thermodynamic functions within the chemical picture. To calculate the ionization/dissociation equilibrium, we use the thermodynamic variational principle: minimization of the free energy with respect to the composition. This approach which is in principle equivalent to the standard method based on Saha equations has some advantages and may lead to additional information in special cases.

Recently, we derived an expression for the free energy of dense hydrogen [18–20] in the framework of the chemical picture. Using Saha equations, the degree of ionization α and the degree of dissociation β as well as the isothermal equation of state (EOS), the hugoniots and the isentropes were calculated. Pauli blocking was taken into account by the concept of excluded volume, which is based on the idea of space occupation by atoms and molecules. Here, it will be shown that a more fundamental approach based on an effective Schrödinger equation leads to important modifications of earlier results. We show here that a more advanced treatment of Paul blocking effects, based on effective Schrödinger equations given in Sect. 3.2, is of essential influence for the theoretical predictions in the region of high densities. The consequences for ionization and dissociation are discussed in Sect. 3.3, and conclusions with respect to the general picture of a transition to highly conducting matter are drawn in Sect. 3.4.

3.2 Bound States in a Plasma

3.2.1 Generalized Beth–Uhlenbeck Equation

The microscopic origin of the concept of excluded volume is the Pauli principle, which is due to the anti-symmetrization of fermionic wave functions. One of the consequences is a contribution to the effective interaction between composite particles and has to be taken into account if a chemical picture is introduced. As an example, the short-range repulsion between atoms or molecules is caused by Pauli blocking of electron orbitals having parallel spin. We focus on the interaction between hydrogen atoms and free electrons. In most of earlier work, the concept of excluded volume in coordinate space has been applied to treat that problem [5, 8–10, 18–20]. Here, we present a microscopic treatment based on the underlying Pauli exclusion principle. The extension to effective interactions between other components of the chemical picture is straightforward, but will not be studied here.

A systematic quantum statistical approach to the equation of state can be given based on the self-energy Σ_{c} and the related spectral function A_{c} ($\mathrm{c}=\mathrm{e},\ \mathrm{p}$), starting, for example, from the normalization condition for the total density of the elementary constituents in terms of the spectral function

$$n_{\mathrm{c}}(\beta,\mu_{\mathrm{e}},\mu_{\mathrm{p}}) = \frac{1}{\Omega}\sum_{1}\int_{-\infty}^{\infty}\frac{\mathrm{d}\omega}{2\pi} f_{\mathrm{c}}(\omega)A_{\mathrm{c}}(1,\omega), \tag{3.1}$$

where $1=\{\boldsymbol{p}_1,\sigma_1\}$ denotes momentum and spin, $f_c(\omega)=[\exp(\beta\omega-\beta\mu_c)+1]^{-1}$ is the Fermi distribution function. We take periodic boundary conditions with respect to the normalization volume Ω, leading to discrete values of the momentum $\boldsymbol{p}$, and the transition from the sum to an integral can be performed as usual in quantum statistics. The spectral function

$$A_{\mathrm{c}}(1,\omega) = \frac{2\mathrm{Im}\,\Sigma_{\mathrm{c}}(1,\omega-\mathrm{i}0)}{(\omega-p_1^2/2m_{\mathrm{c}}-\Sigma_{\mathrm{c}}(1,\omega))^2+(\mathrm{Im}\,\Sigma_{\mathrm{c}}(1,\omega-\mathrm{i}0))^2} \tag{3.2}$$

is related to the self-energy $\Sigma_{\mathrm{c}}(1,z)$ defined in the complex z plane. This way, approximations in the equation of state are traced back to approximations for the self-energy, which can be evaluated at the complex Matsubara frequencies using diagram techniques. Note that thermodynamic potentials are obtained from the equation of state (3.1) by relations such as an expression for the pressure p and the free energy density f.

$$p(T,\mu) = \int_{-\infty}^{\mu} n(\beta,\mu')\,\mathrm{d}\mu', \qquad f(T,n) = \int_0^n \mu(\beta,n')\mathrm{d}n'. \tag{3.3}$$

The chemical picture is rederived if a cluster expansion of the self-energy is performed and the bound-state parts of the few-particle T-matrices occurring in this cluster expansion are taken into account [8]. On the other hand, the low-density expansion of the equation of state is obtained if the two-particle contributions to the self-energy are considered, which in general will contain two-particle bound and scattering states. As long as $\mathrm{Im}\,\Sigma_{\mathrm{c}}(1,\omega - i0)$ can be considered as small quantity, an expansion for the spectral function can be performed, which gives in addition to the quasiparticle δ-like structure also the contribution from the two-particle states. This way, the Beth–Uhlenbeck formula [21] is derived, which provides an exact expression for the second virial coefficient in terms of the two-particle bound state energy and phase shifts.

Our starting point is the generalized Beth–Uhlenbeck formula [5, 21, 22], which relates the densities of the constituent particles in the physical picture to the chemical potentials, but includes medium effects on the mean-field level. In particular, we have for the total electron density with given spin orientation (we mark the spin orientation explicitly, but assume full degeneracy so that $n_{\mathrm{e}}^{\uparrow} = n_{\mathrm{e}}^{\downarrow} = n_{\mathrm{e}}/2$)

$$n_{\mathrm{e}}^{\uparrow}(\beta,\mu_{\mathrm{e}},\mu_{\mathrm{p}}) = \frac{1}{\Omega}\sum_{\mathrm{p}} f_{\mathrm{e}}(E^{\mathrm{qp}}(p)) + \frac{2}{\Omega}\sum_{P,n}^{\mathrm{bound}} g_{\mathrm{ep}}(E_{P,n})$$
$$+\frac{2}{\Omega}\sum_{P,n}\int \frac{\mathrm{d}E}{\pi} g_{\mathrm{ep}}\left(\frac{P^2}{2m_{\mathrm{p}}} + E\right)\frac{\mathrm{d}}{\mathrm{d}E}\left\{\delta_{P,n}(E) - \frac{1}{2}\sin[2\delta_{P,n}(E)]\right\}, \quad (3.4)$$

where $f_{\mathrm{e}}(E^{\mathrm{qp}}(p))$ is the Fermi distribution of electrons with the quasiparticle dispersion relation $E^{\mathrm{qp}}(p) = p^2/(2m_{\mathrm{e}}) + \mathrm{Re}\Sigma_{\mathrm{e}}(p, E^{\mathrm{qp}}(p))$ and $g_{\mathrm{ep}}(E_{P,n}) = [\exp(\beta(E_{P,n} - \mu_{\mathrm{e}} - \mu_{\mathrm{p}})) - 1]^{-1}$ denotes the Bose function for the electron–proton states with total momentum P and internal quantum number n. The spin degrees of freedom are treated explicitly. Disregarding hyperfine splitting into singlet and triplet states, the summation over the spin of the proton gives the factor 2. In general, also the electron–electron interaction channel should be considered, which contributes to the scattering part of the second virial coefficient. In contrast to the simple Beth–Uhlenbeck formula, the free single-particle energy dispersion relation is replaced by the quasiparticle dispersion relation. The two-particle bound state energies $E_{P,n}$ and scattering phase shifts $\delta_{P,n}(E)$ contain medium effects to be discussed in the following subsection. The contribution $\sin[2\delta_{P,n}(E)]$ avoids double counting of contributions, which are already taken into account in the quasiparticle shift.

3.2.2 Effective Schrödinger Equation of Pairs

In the following we use Rydberg units with $m_{\mathrm{e}}/m_{\mathrm{p}} \ll 1, m_{\mathrm{e}} = 1/2, \hbar = 1, e^2/4\pi\epsilon_0 = 2$, so that the binding energies of the isolated hydrogen atom are simply $E_{P,n}^0 = P^2/2m_{\mathrm{p}} + E_n^0$, with $E_n^0 = -1/n^2$, solving the ordinary Schrödinger equation in momentum representation for the electron–proton system after separating the center of mass motion

$$p^2\phi_n(p) - \sum_q V(q)\phi_n(p+q) = E_n^0\phi_n(p). \quad (3.5)$$

With the Coulomb interaction $V(q) = 8\pi/q^2$, we have the normalized wave function for the ground state

$$\phi_1(p) = \frac{8\sqrt{\pi}}{(1+p^2)^2}, \qquad \sum_p |\phi_1(p)|^2 = \int \frac{d^3p}{(2\pi)^3}|\phi_1(p)|^2 = 1. \quad (3.6)$$

In accordance with the normalization conditions, (3.1) and (3.4), which relates the density with the chemical potential, we have to consider a discrete spectrum in momentum space and transform the summation into an integral, and the periodicity volume is taken as $\Omega = 1$.

Embedding the hydrogen atom in a plasma environment, the additional interactions with the medium can be treated within a quantum statistical approach introducing concepts such as the self-energy, dynamical screening, and the spectral function given above. For these many-body quantities, special approximations can be performed, which reflect different processes in the plasma. This way, an effective wave equation has been derived [8,23,24],

$$p^2\psi_n(p) - \sum_q V(q)\psi_n(p+q) + \sum_q H^{\rm pl}(q)\psi_n(p+q) = E_n\psi_n(p). \quad (3.7)$$

The center of mass motion $\boldsymbol{P}$ has been neglected, assuming the adiabatic limit $m_{\rm e}/m_{\rm p} \ll 1$. In general, the plasma Hamiltonian $H^{\rm pl}(q)$ will depend also on $\boldsymbol{P}$ and on the energy if dynamical and retardation effects are taken into account. The plasma Hamiltonian will shift the energy eigen values $E_n = E_n^0 + \Delta E_n$ and will modify the wave functions $\psi_n(p)$. In particular, because of the plasma interaction, the binding energies may merge with the continuum so that the bound states disappear, if the influence of the plasma increases with increasing density. This dissolution of bound states is called Mott effect and has important consequences for the macroscopic properties of the plasma.

We select special contributions to the plasma Hamiltonian, which will modify the bound state properties:

$$H^{\rm pl}(q) = H^{\rm Hartree} + H^{\rm Fock} + H^{\rm Pauli} + H^{\rm MW} + H^{\rm Debye} + H^{\rm polpot} + H^{\rm vdW} + \ldots. \quad (3.8)$$

The first three contributions are of first order with respect to the interaction and determine the mean-field approximation, which is instantaneous in time and contains no dynamic contributions:

$$\begin{aligned}&\sum_q \left[H^{\rm Hartree}(q) + H^{\rm Fock}(q) + H^{\rm Pauli}(q)\right]\psi_n(p+q)\\ &= \sum_{p'} V(0)[2f_{\rm e}(p') - 2f_i(p')]\psi_n(p) - \sum_{p'} V(\mathbf{p}'-\mathbf{p})f_{\rm e}(p')\psi_n(p)\\ &\quad + \sum_{p'} V(\mathbf{p}'-\mathbf{p})f_{\rm e}(p)\psi_n(p'). \end{aligned} \quad (3.9)$$

These contributions are of similar structure and have to be considered simultaneously in consistent approximations. Notice that the electron–electron interaction is repulsive ($V(q)$), and the electron–ion interaction is attractive ($-V(q)$). The Hartree term contains the factor 2 due to spin summation (for abbreviation, only the momentum is given in the Fermi distribution). This contribution vanishes for neutral plasmas where the densities $n_c = \sum_{p'} 2f_c(p')$ compensate each other. The Fock term as well as the Pauli blocking term describe exchange terms and refer only to the interaction between particles of identical species and spin. The origin of the Pauli blocking term is the phase space occupation by free electrons according to the distribution function $f_e(p)$. This phase space cannot be used to form a bound state so that in the atom the interaction of the electron with the ion is blocked when the final state is already occupied by a free electron with same spin orientation.

The following two terms of the plasma Hamiltonian are the Montroll–Ward term giving the dynamical screening of the interaction in the self-energy and the dynamical screening (Debye) of the interaction between the bound particles. These contributions are related to the polarization function and are of particular interest for plasmas due to the long-range character of the Coulomb interaction. In a consistent description, both terms should be treated simultaneously. We will not discuss these terms here in detail, see also [8, 23, 24], and give only some simple estimations below.

The last two contributions to the plasma Hamiltonian are of second order with respect to the interaction. The polarization potential describes the interaction of bound states with free charge carriers, and the van der Waals contribution is due to the interaction between bound states, see [8, 25]. Here, we will not discuss these terms in detail and give only some simple estimations below.

3.2.3 Evaluation of the Mean-Field Energy Shift of Bound States: Perturbation Theory

We first focus on the influence of the mean-field contributions to the effective Schrödinger equation of pairs,

$$E_n\psi_n(p) = p^2\psi_n(p) - \sum_q V(q)\psi_n(p+q) + \sum_q V(q)\left[f_e(p)\psi_n(p+q) - f_e(p+q)\psi_n(p)\right]. \quad (3.10)$$

When the perturbation due to the plasma Hamiltonian is small, the shift of the energy eigen values is obtained with the unperturbed wave functions as

$$E_n - E_n^0 = \Delta E_n^{\mathrm{Fock}} + \Delta E_n^{\mathrm{Pauli}} = -\sum_{p,p'} \phi_n^*(p) V(\mathbf{p}'-\mathbf{p}) f_e(p')\phi_n(p) + \sum_{p,p'} \phi_n^*(p) V(\mathbf{p}'-\mathbf{p}) f_e(p)\phi_n(p'). \quad (3.11)$$

Here, inserting the Schrödinger equation, the Pauli blocking term can be rewritten as

$$\Delta E_n^{\mathrm{Pauli}} = \sum_p \phi_n^*(p)(p^2 - E_n^0) f_{\mathrm{e}}(p)\phi_n(p). \tag{3.12}$$

A simple expression is found in the low-temperature, low-density limit, where the Fermi distribution with the normalization $\sum_p f_{\mathrm{e}}(p) = n_{\mathrm{e}}/2$ is concentrated near $p = 0$. In the zero temperature limit, we have a Fermi sphere with Fermi momentum $p_{\mathrm{F}} = (3\pi^2 n_{\mathrm{e}})^{1/3} \ll 1$. The shift of the ground state $\phi_1(p)$, (3.6), results as

$$\Delta E_1^{\mathrm{Pauli}} = \frac{1}{2} n_{\mathrm{e}} \, (-E_1^0) \, |\phi_1(0)|^2. \tag{3.13}$$

Using Rydberg units, the dimensionless density of free electrons n_{e} is given in units of a_{B}^3, and the ground state energy is $E_1^0 = -1$. The Pauli blocking shift is linear in the density

$$\Delta E_1^{\mathrm{Pauli}} = 32\pi n_{\mathrm{e}}. \tag{3.14}$$

Similarly, the Fock term can be evaluated as

$$\Delta E_1^{\mathrm{Fock}} = -128 \int_0^\infty \mathrm{d}p \frac{1}{(1+p^2)^4} = -20\pi n_{\mathrm{e}}. \tag{3.15}$$

It compensates partially the Pauli shift so that the total shift is

$$\Delta E_1^{\mathrm{Fock}} + \Delta E_1^{\mathrm{Pauli}} = 12\pi n_{\mathrm{e}}, \tag{3.16}$$

which is shown in Fig. 3.1 as dashed–dotted line, indicating a rather steep shift of the bound state energy.

Because of phase space occupation, the bound state energy is shifted and may merge with the continuum of scattering states, indicating the dissolution of bound states. Considering in (3.10) the continuum part of the spectrum describing scattering states, only the Fock shift contributes to the energy shift. The lowest energy in the continuum occurs at $p = 0$ and is shifted by $\Delta E^{\mathrm{Fock}}(p=0) = -\sum_q V(q) f_{\mathrm{e}}(q) = -4p_{\mathrm{F}}/\pi = -4(3n_{\mathrm{e}}/\pi)^{1/3}$. However, the two-particle continuum state can only be created at the Fermi momentum as all states below p_{F} are occupied. Thus the continuum of scattering states begins at p_{F} where we have in the zero temperature limit the Fock shift

$$\Delta E^{\mathrm{Fock}}(p_{\mathrm{F}}) = -\sum_p V(\boldsymbol{p} - \boldsymbol{p}_{\mathrm{F}}) f_{\mathrm{e}}(p) = -2p_{\mathrm{F}}/\pi = -2(3/\pi)^{1/3} n_{\mathrm{e}}^{1/3} \tag{3.17}$$

shown also in Fig. 3.1.

Extrapolating these low-density results to higher densities, the ground state disappears at a density that corresponds in first approximation to $n_{\mathrm{e}} \simeq$

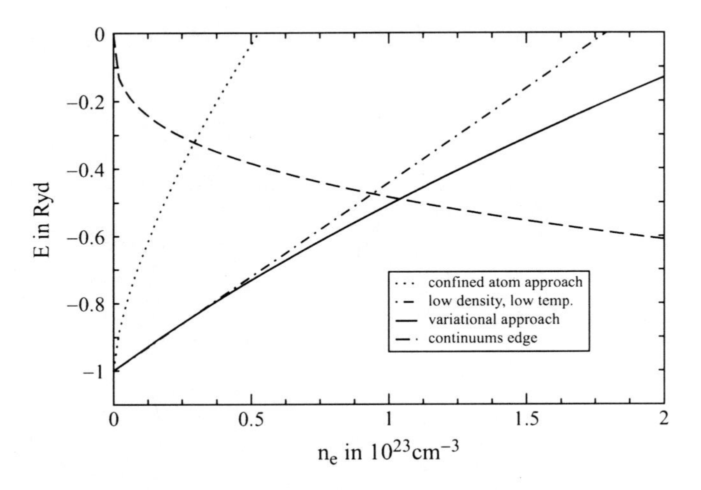

Fig. 3.1. Density dependence of the effective energy of the ground state of hydrogen, low-temperature estimate (*dashed–dotted line*) according to (3.16) in comparison with the confined atom estimate (*dotted line*). The *full line* corresponds to the variational approach. We have shown also the lowering of the continuum edge according to (3.17) (*dashed line*)

$0.015 a_{\mathrm{B}}^{-3} = 10^{23}\ \mathrm{cm}^{-3}$. This leads to an average distance of $r_0 \simeq 2a_{\mathrm{B}}$ and is below the Mott criterion. The Mott condition

$$r_0 \simeq a_{\mathrm{B}}, \qquad \frac{4\pi}{3} n_{\mathrm{e}} r_0^3 = 1 \tag{3.18}$$

expresses the idea that atoms are destroyed if the mean distance of the electrons is equal or smaller than the Bohr radius.

We can also give another discussion of the Mott criterion, which is based on an alternative estimate of the binding energy shift. To compare our expression for the shift, we may take, for example, the confined atom model [8, 26, 27], which assumes that the atom is embedded into a hard sphere with radius r_0. This assumption is based on the idea that the electrons form a kind of hard wall around the atom. In first approximation, this theory gives the shift (in Rydberg units)

$$\Delta E_1^{\mathrm{ca}} = \pi^2 r_0^{-2}. \tag{3.19}$$

Correspondingly, the energy would disappear at $r_0 \simeq 3a_{\mathrm{B}}$, that is, already at a much smaller density (see also Fig. 3.1). Better estimates based on numerical solutions of the Schrödinger equation give a value of about $r_0 \simeq 2a_{\mathrm{B}}$. Our first estimate is of same order.

The confined atom model is closely related to the concept of excluded volume, which assumes for each of the components a particular volume that is not available for the other particles of the same species, so that the Pauli blocking mechanism is introduced in an elementary way. At the same time, we also have a shift of the free electron energies due to the Pauli blocking by electrons bound in atoms. Within the cluster mean-field approximation [8,28] these shifts are given by self-energy shift due to bound states

$$\Delta E^{\text{Pauli bl.}}_{\text{free electr.}}(p) = \sum_{2,3} V(12,12)|_{\text{ex}} g(E_1)|\psi_1(2,3)|^2 - \sum_{2,1',2'} g(E_1)V(12,1'2')\psi_1(12)\psi_1^*(1'2'). \quad (3.20)$$

The Hartree contributions vanish because of charge neutrality, but the exchange contributions produce a shift of the single-electron energies.

It is possible to evaluate the Pauli blocking shift of the bound states for arbitrary temperatures within perturbation theory. In the general case of arbitrary temperatures and densities, the Fermi function cannot be simplified and we have to evaluate the full integral numerically

$$\Delta E_1^{\text{Pauli}} = \frac{32}{\pi}\int_0^\infty \mathrm{d}p \frac{p^2}{(1+p^2)^3} f_{\mathrm{e}}(p). \quad (3.21)$$

If we approximate the Fermi distribution by a Boltzmann distribution normalized to the same density, we obtain an analytical expression

$$\Delta E_1^{\text{Pauli}} \approx 32\pi n_{\mathrm{e}} G(T/T_0), \quad (3.22)$$

where the function $G(T/T_0)$ expressing the temperature-dependence is given by

$$G(x) = \frac{1}{x^{7/2}}\left[\sqrt{x}\left(1+\frac{x}{2}\right) - \sqrt{\pi}\left(1 - x - \frac{x^2}{4}\right)\exp\left(\frac{1}{x}\right)\left(1 - \mathrm{Erf}\left[\frac{1}{\sqrt{x}}\right]\right)\right] \simeq \frac{1}{1+77\,x/16}. \quad (3.23)$$

Here $T_0 = 1\,\mathrm{Ryd}/k_{\mathrm{B}} = 157,886\,\mathrm{K}$ is the ionization temperature. In the asymptotic approximation $T/T_0 \ll 1$ (i.e., for temperatures below 20,000 K where $G(x) \simeq 1$) this leads back to the zero temperature expression for the shift given above. A similar expression can also be given for the Fock term.

3.2.4 Evaluation of the Mean-Field Energy Shift of Bound States: Variational Approach

According to our estimate, the effective binding energy would disappear at $n_{\mathrm{e}} \simeq 0.01 a_{\mathrm{B}}^{-3}$. This is clearly too early, that is, at too low densities. The reason is that perturbation theory tends to overestimate effects, that is, the shifts are rather steep and the extrapolation up to the Mott density is questionable. For the evaluation of the bound state energy over a large region of density and temperature, we have to modify the wave function that will be performed within a variational approach.

In particular, we are interested in the Mott effect, describing the dissolution of the bound state at increasing density when the bound state energy merges with the continuum of scattering states. The Fock shift acts for free states and bound states. We assume that the wave function changes smoothly from the bound state to the continuum state near the Mott density. Consequently, the Fock shift of the bound state and the continuum states become identical there, and the bound state wave function becomes indistinguishable from the continuum states. Then, the binding energy, which is the difference between the bound state energy and the continuum of scattering states, is not depending on the Fock shift near the Mott density. To obtain the Mott density, we can restrict us first to the Pauli blocking contribution.

To apply the Ritz variational approach, we have to symmetrize the Hamiltonian in the effective wave equation (3.24) introducing the function $\Psi_n(p) = \psi_n(p)[1 - f_e(p)]^{-1/2}$

$$[p^2 + \Delta^{\mathrm{Fock}}_{\mathrm{free\,electr.}}(p)]\Psi_n(p) - \sum_{p'}[1 - f_e(p)]^{1/2}V(p-p')[1 - f_e(p')]^{1/2}\Psi_n(p') = E_n\Psi_n(p). \tag{3.24}$$

We consider the zero temperature case and use the *ansatz* corresponding to a variable Bohr radius,

$$\Psi_0(p;\alpha) = \frac{8\pi^{1/2}}{\alpha^{3/2}}\left(1 + \frac{p^2}{\alpha^2}\right)^{-2}. \tag{3.25}$$

In a more refined approach we take into account that no states below the Fermi momentum are available to build the bound state,

$$\Psi(p;\alpha) = \frac{1}{N}\frac{1}{(1 + p^2/\alpha^2)^2}\Theta(p - p_\mathrm{F}). \tag{3.26}$$

Here α is a parameter that characterizes the occupation in the momentum space. Our ansatz gives the ground state energy including the Pauli shift in the zero temperature limit. With $f = p_\mathrm{F}/\alpha$, the energy is calculated as the sum of kinetic and potential energy,

$$\begin{aligned}
E(p_\mathrm{F},\alpha) &= \alpha^2\left[1 + 32\, f^3/N_1\right] - 6\alpha(Z_1 + Z_2)/N_2\\
N_1 &= \pi\left[2f(3 - 8f^2 - 3f^4) + 3(1+f^2)^3\pi - 6(1+f^2)^3\arctan(f)\right],\\
Z_1 &= (1+f^2)[(1+f^2)^2\pi^2 - 4f(f^2-1)\pi + 4f^2],\\
Z_2 &= (1+f^2)\{4\arctan(f)[2f(f^2-1) - (1+f^2)^2\pi + (1+f^2)^2\arctan(f)]\},\\
N_2 &= \pi[3\pi(1+f^2)^3 - 6(1+f^2)^3\arctan(f) + 2f(3 - 8f^2 - 3f^4)].
\end{aligned} \tag{3.27}$$

For given $p_\mathrm{F} = (3\pi^3 n_\mathrm{e})^{1/3}$, the minimum with respect to α yields an estimate for the ground state energy. As self-energy shifts are irrelevant for the Mott condition, the bound states are dissolved when the ground state energy becomes zero. The evaluation of the binding energy (3.27) gives for the Mott condition $n_e = 0.033$, $\alpha = 0.77$. The variational ansatz (3.26) can be improved with $\Psi(p;\alpha) = N^{-1}(1 + (p - p_F)^2/\alpha^2)^{-2}\Theta(p - p_F)$ which gives a better transition to the continuum states.

For exploratory calculations, we take the wave function (3.25) and change the parameter α. First we estimate the energy shifts by the wave function at zero momentum as in perturbation theory in the zero temperature, low density case, which gives for the ground state energy

$$E_1 \approx \min_\alpha E_1(\alpha) = \min_\alpha(\alpha^2 - 2\alpha + [4\pi + (128/3)]n_e/\alpha^2). \tag{3.28}$$

The shift of energy is in good agreement with the result obtained from (3.27). As examples we show the energy curves for $n_e = 0$, $n_e = 0.005a_B^{-3}$, and $n_e = 0.0305a_B^{-3}$ in Fig. 3.2, the latter density corresponding just to the disappearance of the ground state energy. The shift of the binding energy as function of the density is also shown in Fig. 3.1. Within a better approximation, we take the Fermi function in the zero temperature limit, $f_e(p) = \Theta(p_F - p)$, and evaluate the Pauli blocking shift integrating over the wave function $\Psi_0(p;\alpha)$,

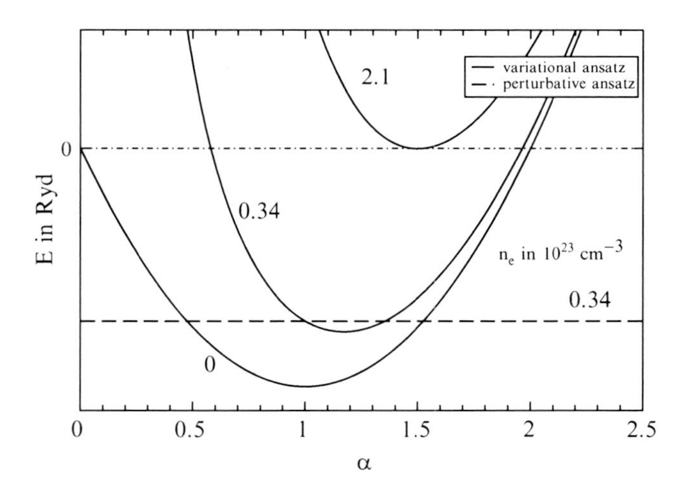

Fig. 3.2. Parameter dependence of the effective energy of the ground state including shifts in Boltzmann approximation, calculated in variational approach, for different electron densities. The minima correspond to the appropriate values of α. The zero density corresponds to an isolated hydrogen atom. At the intermediate density, the variational approach is already much lower than the perturbation theory (*dashed line*). The highest density corresponds to situation where – within the present approximation – the ground state merges into the continuum

$$\Delta E_1^{\text{Pauli}} \approx \frac{4\pi}{(2\pi)^3} \int_0^{p_\text{F}} \mathrm{d}p p^2 \left[\frac{p^2}{\alpha^2} + 1\right] \psi_0^2(p)$$
$$= \frac{4\alpha^2}{\pi} \left[\frac{\frac{p_\text{F}}{\alpha}\left(\frac{p_\text{F}^2}{\alpha^2} - 1\right)}{\left(1 + \frac{p_\text{F}^2}{\alpha^2}\right)^2} + \arctan\left(\frac{p_\text{F}}{\alpha}\right)\right], \tag{3.29}$$

with the Fermi momentum ${p_\text{F}}^3 = 3\pi^2 n_\text{e}^*$ (all momenta are given in Rydberg units).

We can also calculate the temperature dependence of the Pauli blocking term. For this, we have to replace the zero-temperature Fermi function in the interaction term by the finite temperature distribution. It can be seen that the temperature dependence of the Pauli blocking term becomes week. Even within the variational approach, the densities where the energy levels disappear and consequently full ionization occurs are evidently still too low to explain the observed effects.

3.2.5 Evaluation of the Mean-Field Energy Shift of Bound States Including the Fock Term

The Fock term occurs in the Bethe–Salpeter equation as mean-field contribution and is of the same order as the Pauli blocking term. Even if the Fock term is not of primary importance for the disappearance of the bound state energy, it has to be included in the total shift of bound and scattering states to be consistent (so-called conserving approximations). Within perturbation theory, in the zero temperature limit, we get after some transformations the integral

$$\Delta E_1^{\text{Fock}} = -\frac{64}{\pi^2} \int_0^\infty \frac{p\,\mathrm{d}p}{(1+p^2)^4} \int_0^{p_F} k\,\mathrm{d}k \, \ln\frac{(p+k)}{|p-k|}$$
$$= -\frac{64}{\pi^2} \int_0^\infty \frac{p\,\mathrm{d}p}{(1+p^2)^4} \left[p_\text{f} p + \frac{1}{2}(p_\text{F}^2 - p^2) \ln\frac{(p+p_\text{F})}{|p-p_\text{F}|}\right]$$
$$= -\frac{4}{3\pi} p_\text{F}^3 \frac{5+3p_\text{F}^2}{(1+p_\text{F}^2)^2}, \tag{3.30}$$

which reproduces in the low-density limit the value $(-20\pi n_\text{e})$ given above in (3.15).

By summing up the linear Pauli and Fock contributions to the density expansion of the shift of bound states, we find in the low-density limit a positive term as shown above. Assuming that the temperature dependence of the Fock shift is the same as that for the Pauli shift, we may define a temperature dependent coefficient of the linear shifts by

$$\Delta E_1^{\text{lin}} \simeq a'(T) n_\text{e} \text{ with } a'(T) = 12\pi G\left(\frac{T}{T_0}\right). \tag{3.31}$$

For convenience of the numerical procedure in the later variational calculations of the free energy, we constructed an interpolation formula between the Boltzmann and the zero temperature limits, (3.29) and (3.30), by taking into account some points of numerical evaluations. We took the basic structure of the asymptotes given by (3.29) and (3.30) and made a minimum of changes to get the right steepness at $n_{\mathrm{e}} = 0$. As a result, we are proposing for the Pauli shift the interpolation formula

$$\Delta E_1^{\mathrm{Pauli}} = \frac{4}{\pi}\left[\frac{p_{\mathrm{F}}(c(T)p_{\mathrm{F}}^2 - 1)}{(1 + c(T)p_{\mathrm{F}}^2)(1 + p_{\mathrm{F}}^2)} + \arctan(p_{\mathrm{F}})\right]. \tag{3.32}$$

This is nearly identical to the asymptotic representation, except that we had to introduce a fit function

$$c(T) = \frac{1}{3}\left(G(T) - 1\right) \tag{3.33}$$

to provide the correct steepness at small densities. The Fock term is less easy. Following the same strategy as above, we derive the interpolation formula

$$\Delta E_1^{\mathrm{Fock}} = -\frac{20\pi}{g}\ln\left(1 + g n_{\mathrm{e}} + k n_{\mathrm{e}}^2 + l n_{\mathrm{e}}^3\right) \tag{3.34}$$

with the fit parameters $g = 261.65$, $k = 60{,}000$, and $l = 334{,}369$.

A comparison of the density dependence according to the interpolations introduced above with numerical estimates of the integrals is shown in Fig. 3.3

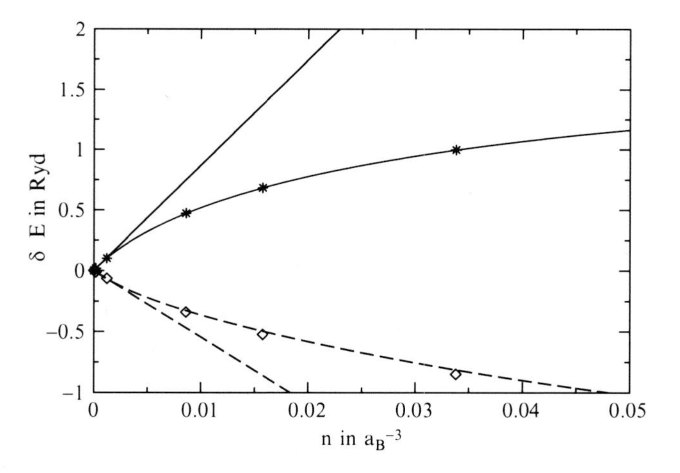

Fig. 3.3. Results for the Pauli shift (*full lines, stars*) and the Fock shift (*dashed lines, diamonds*) for the temperature $T = 5{,}000$ K. The interpolation formulae, based on the asymptotic representations and the numerical evaluations (*stars, diamonds*), are shown in comparison to the Boltzmann approximation (the cone defined by the outer *straight lines*)

for $T = 5{,}000\,\mathrm{K}$. We see that the agreement with the data is for this temperature quite reasonable. We mention that the temperature dependence in the region of interest $5{,}000 < T[\mathrm{K}] < 15{,}000$ is quite weak.

3.2.6 Discussion of Further Contributions to the Shift

The remaining shifts are smaller and will be discussed here only qualitatively on the basis of the data from literature. Let us first discuss the shifts connected with screening effects. As known for the ground state of hydrogen, the shifts due to screening in the self-energy and in the effective interaction compensate in the leading order $n_e^{1/2}$ of density, in contrast to the scattering states where only the self-energy acts. A detailed discussion of the contributions of dynamical screening $H^{\mathrm{MW}}(q) + H^{\mathrm{Debye}}(q)$ to the plasma Hamiltonian (3.8) has been given elsewhere, see [23, 29]. From the Bethe–Salpeter equation, we find by using perturbation theory

$$\Delta_1^{\mathrm{scr}} \simeq -(Z-1)^2 \frac{e^2}{2r_D} - Bn_{\mathrm{e}} + \cdot, \tag{3.35}$$

where r_{D} is the Debye screening radius. For hydrogen the first term proportional to the root of the density disappears as $Z = 1$. These terms are rather complicated and not completely known analytically up to now. This may change in higher orders: it is known that the ground state energy decreases slightly with the density. Inspecting the illustrations in [8], we find approximately $B \simeq 20$–30 for all shifts together. The more recent calculations by Arndt et al. [30] which include dynamical effects give a smaller value $B \simeq 10$ for the sum of all shifts. However, no specific results on the Pauli and Fock shifts are given in these studies. Therefore, a more detailed comparison is difficult at present time. We rely on our own calculations and neglect here the screening shifts. However, the shifts due to screening might be of the same order as the shift due to polarization.

In a similar way as we have calculated the shifts discussed above, we may also calculate the shift due to polarization of the atoms by free charges [25]. Within a semiclassical approach neglecting degeneracy we have in Rydberg units

$$\Delta E_1^{\mathrm{polpot}} = -\frac{27}{2}\frac{n_{\mathrm{e}}}{R_{\mathrm{ae}}}\frac{1}{1 + f_1^2 (R_{\mathrm{ae}}/r_{\mathrm{D}})^2}, \tag{3.36}$$

where R_{ae} is an effective minimal electron–atom distance and $\kappa = 1/r_{\mathrm{D}}$ with a numerical factor f_1. As an estimate we use $f_1 = 2/3$. The shift due to polarization effects, expressed by (3.36), leads to negative contributions to the energy shifts and correspondingly to the free energy density. According to these estimates, the contributions due to polarization might reduce the Pauli blocking and Fock effects by about 20%. However, the Pauli blocking shift remains the largest among all shifts. Therefore, we assume in the following that the Pauli blocking shift determines the temperature and the density dependence.

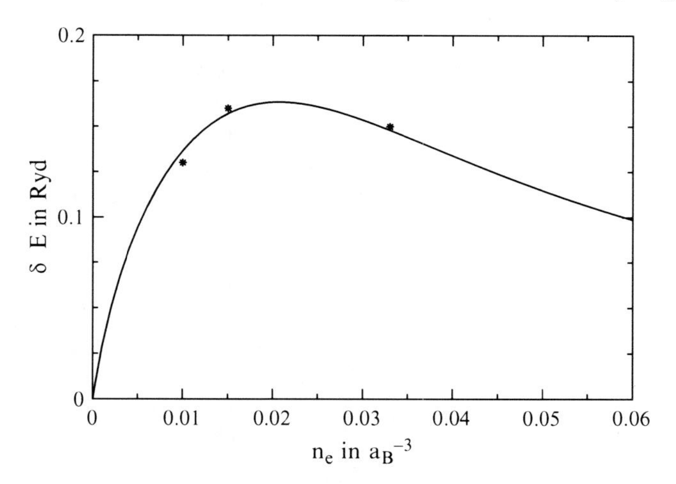

Fig. 3.4. Dependence of the total energy shift of the ground state of hydrogen including Pauli and Fock contributions, according to (3.31), on the electron density n_e. The interpolation formula (*full line*) connects several points from the numerical evaluation (*stars*)

In conclusion, we may say that in the region of interest the low-density limits of the Pauli blocking and the Fock contribution are of special relevance. We take both dominant terms into account. The polarization term that might give a correction of about 20%, in the worst case scenario, was neglected in the following calculations of the ionization equilibrium. The self-energy and screening shifts (the terms proportional to $V^{\mathrm{eff}} - V$ in the effective wave equation), which might give additional corrections of about 10%, were also neglected here. Another correction that possibly may shift the Mott density to higher values is the improvement of the mean-field approximation, including T-matrix contributions to the self-energy. Correlations between the atom and the free electrons will reduce the Pauli blocking term.

3.3 Thermodynamic Functions and Ionization Equilibrium of Hydrogen

3.3.1 The Chemical Picture

We construct the thermodynamic functions of hydrogen by using a chemical approach to the free energy, which recently was applied to temperatures between 2,000 and 10,000 K [18,19]. The effects of pressure dissociation, $H_2 \rightleftharpoons 2H$, and ionization, $H \rightleftharpoons e + p$, are taken into account so that the transition from a molecular fluid at low temperatures and pressures through a partly dissociated, warm fluid at medium temperatures of some thousand Kelvin to a fully ionized, hot plasma above $10,000\,\mathrm{K}$ can be explained.

Before we start with a discussion of the various contributions to the free energy $F(T,V,N)$, we have to say that any splitting of F is conditional. Depending on the picture we use, the contributions may change, but the total free energy F should be invariant with respect to all possible models. Because of the strong statement, we have to make sure that no physical effect, including in particular the Pauli effects which we want to study here, should be accounted for twice. We check this very carefully.

In a chemical picture, hydrogen consists basically of two main components, the plasma (electrons and protons) and the neutral fluid consisting of composite particles or bound states (atoms and molecules). Correspondingly, the free energy expression for a two-component system of neutral ($F_{\rm nl}$) and charged particles ($F_{\rm pl}$) reads

$$F(V,T,N) = F_{\rm bs} + F_{\rm nl} + F_{\rm pl} + F_{\rm int}. \tag{3.37}$$

The bound state contribution will be written here in the form

$$F_{\rm bs} = N_a^* E_0^a + N_m^* E_0^m + N_a^* k_{\rm B}T \,\ln \sigma'(T) + N_m^* \,\ln \sigma_m'(T), \tag{3.38}$$

where the reduced atomic partition function of Brillouin–Planck–Larkin (excluding the ground state) is given by

$$\sigma'(T) = 4\,\exp(\beta E_0^a) \sum_{n>0} \left[\exp(-\beta E_n) - 1 + \beta E_n\right]. \tag{3.39}$$

The molecular partition function has been discussed elsewhere [18]. Beyond the bound state contributions, we have the usual free energy contributions of the neutral gas $F_{\rm nl}$ and the plasma $F_{\rm pl}$. The latter contributions are split into an ideal and an interaction part and finally we have a contribution of the interactions between the neutrals and the charges $F_{\rm int}$. We remark that the splitting into different terms is conditional within the chemical picture. We have some freedom in splitting, but in any case double counting of contributions has to be strictly avoided.

Basically, we concentrate here on the contributions of the energy shifts. We approximate the shift of the atomic levels by the sum of Pauli–Fock terms and polarization terms

$$\Delta E_0^a = \Delta E_0^{\rm Pauli-Fock} + \Delta E_0^{\rm polpot}. \tag{3.40}$$

We assume that molecules are simply composites of two atoms, that is, the shifts are additive:

$$\Delta E_0^m = 2\Delta E_0^a. \tag{3.41}$$

Concentrating in this work on the influence of the atomic shifts, we will not go into the details of the other contributions due to the neutrals and the charged particles to the free energy, just making a few remarks. The plasma term combines several results for the fully ionized plasma domain in the form of a Padé approximation [6, 7, 31].

The neutral gas contribution is given by expressions based on improved data for the dense, neutral fluid calculated within a dissociation model [32, 33]. In earlier work, we have performed classical Monte Carlo simulations for partially dissociated, fluid hydrogen for a grid of temperature and density points in the region of $T = (2 - 10) \times 10^3\,\mathrm{K}$ and $\varrho = (0.2 - 1.1)\,\mathrm{g\,cm^{-3}}$. Effective pair potentials of the exponential-6 form have been used to model the interactions between the molecules and the atoms in the dense fluid. The dissociation equilibrium $H_2 \rightleftharpoons 2\,H$ has been solved taking into account the correlation parts of the chemical potentials using fluid variational theory. The Monte Carlo data for the interaction contribution can be interpolated accurately within an eight-parameter fit with respect to density and temperature, leading to an analytical expression for the free-energy density, see [18, 19, 34]. The resulting pressure of the neutral component may be presented as

$$p_{\mathrm{nl}} = \frac{1}{2}(1 + \beta) n k_{\mathrm{B}} T + p_{\mathrm{int}}. \tag{3.42}$$

Here n is the total proton density and β is the degree of dissociation of molecules into atoms. The interaction part can be fitted by the following polynom [19],

$$\begin{aligned} p_{\mathrm{int}}[\mathrm{GPa}] &= \rho^2[2055 - 2469\rho^2 + 547.2\rho^3 - 3.351\sqrt{T} + 3.882\rho\sqrt{T}] \\ &\quad + \rho^2 \ln\rho[688.2 + 2395\rho - 1.986\sqrt{T}], \end{aligned} \tag{3.43}$$

where the temperature is given in K and the density in $\mathrm{g\,cm^{-3}}$, the pressure is given in GPa. The free energy follows by integration. The degree of dissociation is obtained by solving the (ideal) mass action law, which reads

$$\beta = \frac{1}{nK_{\mathrm{m}}}\left[(1 - 4nK_{\mathrm{m}})^{1/2} - 1\right], \tag{3.44}$$

where K_{m} is the mass action constant of the atom–molecule equilibrium. As the fit (3.43) is quite soft, we have used the Carnahan–Starling expression for hard spheres [35] at temperatures higher than 15,000 K, with a soft transition from the soft fit (3.43) to the relatively hard Carnahan–Starling expression.

The plasma contributions consist also of an ideal and an interaction term. For example, we have for the free energy

$$F_{\mathrm{pl}} = F_{\mathrm{pl,id}} + F_{\mathrm{pl,int}}. \tag{3.45}$$

Taking full account of Pauli blocking effects, we write for the ideal term

$$F_{\mathrm{pl,id}}(V, T, N) = N_{\mathrm{e}} k_{\mathrm{B}} T z \left(\frac{N_{\mathrm{e}} \Lambda_{\mathrm{e}}^3}{2V^*}\right). \tag{3.46}$$

Here V^* is the volume available for the electrons. Basically, considering no excluded volume effects, this is the total volume of the system, that is, $V^* = V$.

Further, $z(y)$ is the Fermi degeneracy function, which we approximate by a formula constructed by Zimmermann [36]

$$z(y) = \ln y - 1 + 0.1768y - 0.00165y^2 + 0.000031y^3, \text{ if } y < 5.5, \quad (3.47)$$
$$z(y) = 0.7254y^{2/3} - 2.0409y^{-2/3} + 0.85y^{-2}, \text{ if } y > 5.5. \quad (3.48)$$

The interaction part of the plasma component is represented by the standard Padé approximations developed in earlier work [6, 7, 31, 36]. We have calculated the chemical equilibrium between the charges and the neutrals (atoms and molecules) by means of a numerical variational procedure based on direct minimization of the free energy. We have used a program based on MATHEMATICA, which was presented in [37]. We show that Pauli blocking has a strong influence on the thermodynamic properties. The present study is based on the energy shifts, which is an alternative approach to the excluded volume concept used in earlier work [18].

3.3.2 The Ionization Equilibrium

We consider a hydrogen plasma at fixed temperature T and proton density n. We take into account ionization $\mathrm{H} \rightleftharpoons \mathrm{p} + \mathrm{e}$ and dissociation processes $\mathrm{H}_2 \rightleftharpoons 2\mathrm{H}$. For simplicity, the formation of other species such as H_2^+ and H^- will be neglected. The degrees of ionization and dissociation are defined by the following expressions [18]:

$$\alpha = \frac{n_i}{n_i + n_\mathrm{a} + 2n_\mathrm{m}},$$
$$\beta_\mathrm{d} = \frac{n_\mathrm{a}}{n_\mathrm{a} + 2n_\mathrm{m}}, \quad \beta_\mathrm{a} = \frac{n_\mathrm{a}}{n_i + n_\mathrm{a} + 2n_\mathrm{m}}, \quad \beta_\mathrm{m} = \frac{2n_\mathrm{m}}{n_i + n_\mathrm{a} + 2n_\mathrm{m}}. \quad (3.49)$$

The free energy has to be minimized with respect to these parameters within a variational procedure. We note that β_d is the degree of dissociation of molecules into atoms, β_a is the relative amount of protons bound in atoms, and β_m is the relative amount bound in molecules. Because of the balance relation for the total proton density

$$n = n_i + n_\mathrm{a} + 2n_\mathrm{m}, \quad (3.50)$$

we find the useful relations

$$\beta_\mathrm{a} = \alpha(1 - \beta_\mathrm{d}), \; \beta_\mathrm{m} = (1 - \alpha)(1 - \beta_\mathrm{d}), \; \beta_\mathrm{d} = \frac{\beta_\mathrm{a}}{\beta_\mathrm{a} + \beta_\mathrm{m}}. \quad (3.51)$$

In other words, only one of the parameters β is independent. We prefer here to use α and β_m and will make use of the simplex relation, $\alpha + \beta_\mathrm{a} + \beta_\mathrm{m} = 1$. It can be shown that atoms appear only in a rather narrow region of the density–temperature plane. In large regions of the density–temperature plane we can assume $\beta_{\mathrm{a}} = 0$, so that α remains as the only free parameter. The condition of neutrality requires that electron and ion densities are always equal, that is, $n_{\mathrm{e}} = n_i = n$.

We introduce the free energy per proton measured in units of $k_{\mathrm{B}}T$ by

$$\phi\,(T, n; x, y) = \frac{F}{k_{\mathrm{B}} T N_{\mathrm{p}}}. \tag{3.52}$$

This basic quantity depends only on four independent parameters: temperature T and density n are given; the variational parameters are the degrees of ionization and dissociation. We denote from now on the degree of ionization by x and the degree of dissociation by y. The equilibrium composition follows from the minimization procedure

$$\frac{\delta\phi}{\delta x} = 0, \qquad \frac{\delta\phi}{\delta y} = 0, \tag{3.53}$$

which yields the real degrees of ionization and dissociation

$$\alpha = x_{\min}, \, \beta_k = y_{\min}. \tag{3.54}$$

Here β_k stands for one of the three dissociation parameters introduced earlier, and we are free in this choice. The density dependence of the degrees is represented in Fig. 3.5. We mention that the function $\phi\,(n, T; x, y)$ provides us not only with the physical value of the degrees that correspond to the minima, but also with a dispersion around the minima [37] based on Onsager-type relations.

We calculated the chemical equilibrium by means of a numerical variational procedure based on direct minimization of the free energy. The calculus based on MATHEMATICA was first presented in [37]. We prefer here the minimization of the free energy in comparison with the Saha approach because of several numerical advantages. The results of the two approaches (minimization and Saha approach) are similar in qualitative respect but differ in details. Saha equations are a more explicit formulation of the minimum condition. The Saha equation finds one concrete minimum located inside the corner of the ionization/dissociation degrees. Minima at the boundaries, say $\alpha = 1$ or $\beta = 0$, cannot be found. The direct minimization of the free energy used here finds all existing minima, including multiple minima and those at the boundaries. Further we mention that the iterative solution of Saha equations, which are highly nonlinear, may lead to serious numerical instabilities and convergence to spurious solutions. In particular, these difficulties may appear if the Saha equation contains mixed nonlinearities with respect to electronic and atomic densities in the exponent. Just this is the case for our Pauli blocking effects. Therefore, we may consider the direct minimization of the free energy as a more reliable method. It provides more information and contains the solutions of the Saha equation as special cases.

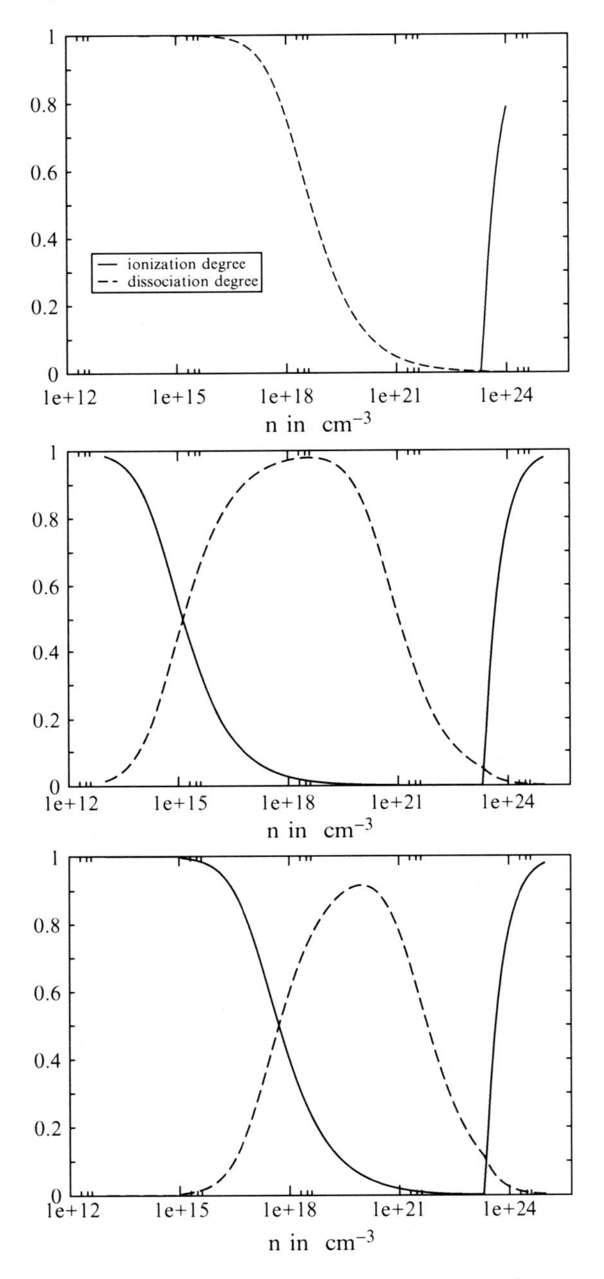

Fig. 3.5. Degree of ionization and degree of dissociation for temperatures T of 5,000 K (*top*), 10,000 K (middle), and 15,000 K (*bottom*) as function of the total proton density (taking into account the energy shifts from Sect. 1 and Padé approximation without polarization)

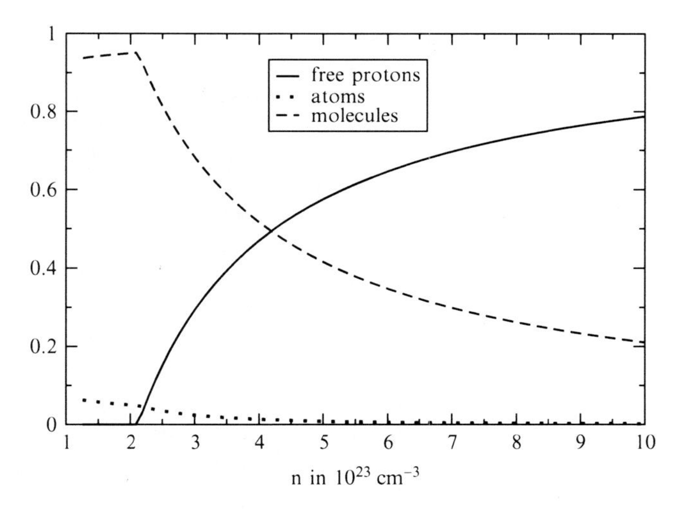

Fig. 3.6. Degree of free protons and of protons bound in molecules at $T = 10{,}000\,\mathrm{K}$ for the transition region of the density. The percentage of protons bound in atoms follows as the difference to 1 (simplex condition)

The transition density is between 10^{23} and 10^{24} protons cm^{-3}. This means that we are in the region around $0.5\,\mathrm{g\,cm^{-3}}$ for hydrogen and $1.0\,\mathrm{g\,cm^{-3}}$ for deuterium. At 5,000 K the transition occurs in the region of density around $n = 3.0 \times 10^{23}\,\mathrm{cm}^{-3}$. To illustrate what happens during this transition, we have shown the behavior of α and β_{a} (percentage of protons bound in atoms) in Fig. 3.6. We see that the transition to full ionization is rather soft.

In principle, all thermodynamic functions may be calculated from the free energy (3.37) by standard thermodynamic relations. For instance, the pressure (isothermal EOS) and the entropy follow from

$$p(T, N/V) = -\frac{\partial F(V,T,N)}{\partial V}, \tag{3.55}$$

$$S(T, V, N) = -\frac{\partial F(V,T,N)}{\partial T}. \tag{3.56}$$

Combining these two expressions we may get the usual EOS $p = p(\rho, T)$ as well as the isentropic (adiabatic) EOS $p = p(s, T)$, where ρ is the total mass density and $s = S/Nk_{\mathrm{B}}$ the specific entropy per proton ($N = N_{\mathrm{p}}$ – total number of protons in the plasma including the protons bound in H and in H_2). In earlier work, we have calculated the isentropic EOS based on the chemical picture [34, 38].

3.4 Discussion and Conclusions

We contribute here to the theory of hydrogen at high pressures in the region where a Mott transition has been predicted and where recent experiments have shown a transition from insulating behavior to metal-like conductivity [39–48]. To understand this transition, several effects have to be taken into account. We concentrated here on so-called Pauli blocking effects expressing the rule that states occupied by atomic electrons cannot be occupied by free electrons with the same spin direction. This leads at high electron densities to the destruction of atomic states, which need a relatively high amount of phase space. We calculated the energy shifts due to Pauli and Fock effects. On this basis we discuss the Mott transition by solving effective Schrödinger equations for strongly correlated systems.

The ionization and dissociation equilibria are treated within an advanced chemical approach based on the assumption that the system is a gaseous mixture of chemical species. The theory for the components is based on expressions for the free energy developed recently to determine Hugoniot curves and isentropes in dense hydrogen plasmas in the regions of partial dissociation and partial ionization. We have shown here that the effects resulting from the Pauli exclusion of electrons from the interior of atoms has a major influence in the high pressure region. We presented explicit calculations of the ionization and dissociation equilibria in the region $5{,}000 < T[\mathrm{K}] < 15{,}000$, $0.1 < \rho[\mathrm{g\,cm^{-3}}] < 1$. At higher temperatures we observe a transition from a neutral hydrogen gas to a highly ionized plasma.

The standard chemical model was improved in the present work in several respects. We replaced the standard excluded volume approximation used in our earlier work by a more rigorous approach based on effective wave equations including symmetry effects, which were solved in several approximations including perturbation theory and variational approaches. This allowed us to include the interactions between electrons and neutrals in a more systematic way. We calculated the shift of the ground state energy due to the effects of Pauli exclusion, which prevent the electrons from penetrating into atoms [49].

This work concentrated primarily on hydrogen. To use the hydrogen EOS for deuterium, mass scaling can be applied for the interpolation formula of the interaction contributions, that is, it is assumed that the same particle numbers for hydrogen and deuterium lead to the same degree of dissociation and to identical interaction contributions to the thermodynamic function of the neutral fluid for a given temperature. Let us conclude with a methodological remark: This work is based on the chemical picture. As well known, there exist two variants of the chemical picture:

1. The calculations of the chemical equilibrium are based on the concept of Saha-type equations, which are reformulations of the equilibrium conditions for the chemical potentials of the species.
2. Alternatively, one can use a direct minimization of the free energy with respect to the variations of the degrees of ionization/dissociation.

Both methods are based on the same variational principle for the free energy of the plasma. The results are similar in qualitative respect but differ in details. Saha equations are a more explicit formulation of the minimum condition. The Saha equation finds one concrete minimum located inside the corner of the ionization/dissipation degrees. Minima at the boundaries, say $\alpha, \beta = 0$ or $\alpha, \beta = 1$, cannot be found. The direct minimization of the free energy used here finds all possible minima, including relative minima and those at the boundaries. Therefore, we may consider the direct minimization of the free energy used here as a more advanced description. It provides more information and contains the solutions of the Saha equation as special cases [50].

Let us summarize the main physical results obtained in this work. The Pauli blocking effects are quite essential for the ionization equilibria and determine the transition to highly conducting states, which occur at densities around 3×10^{23} protons cm^{-3}, in the region of $0.5\,\mathrm{g\,cm}^{-3}$ for hydrogen and $1\,\mathrm{g\,cm}^{-3}$ for deuterium. The corresponding pressures are in the region of $0.8 - 1.2 \times 10^{11}\,\mathrm{Pa}$, that is, around 1 Mbar ($10^{11}\,\mathrm{Pa} = 1\,\mathrm{Mbar}$).

Acknowledgment

We acknowledge helpful discussions with Wolf D. Kraeft. He also performed a detailed comparison with his earlier analytical and numerical results.

References

1. E. P. Wigner, H.B. Huntington, J. Chem. Phys. **3**, 764 (1935)
2. A.A Abrikosov, L.P. Gorkov, L.P. Dsyaloshinsky, *Methods of Quantum Field Theory in Statistical Physics* (in Russian, Moscow, 1962)
3. G.I. Kerley, Los Alamos Scientific Laboratory Report, LA–4776, January 1972
4. E.G. Brovman, Yu. Kagan, A. Kholas, Fiz. Tverd. Tela **12**, 1001 (1970); E.G. Brovman, Yu. Kagan, Usp. Fiz. Nauk **112**, 369 (1974)
5. W. Ebeling, W.D. Kraeft, D. Kremp, *Theory of Bound States and Ionization Equilibrium* (Akademie Verlag, Berlin, 1976)
6. W. Ebeling, Physica **130 A**, 587 (1985)
7. W. Ebeling, W. Richert, Phys. Stat. Sol. B **128**, 467 (1985); Phys. Lett. A **108**, 80 (1985); Contrib. Plasma Phys. **25**, 1 (1985)
8. W.D. Kraeft, D. Kremp, W. Ebeling, G. Röpke, *Quantum Statistics of Charged Particle Systems* (Plenum Press, New York, 1986)
9. D. Saumon, G. Chabrier, Phys. Rev. Lett. **62**, 2397 (1989); Phys. Rev A **46**, 2084 (1992)
10. D. Kremp, M. Schlanges, W.D. Kraeft, *Quantum statistics of nonideal plasmas* (Springer, Berlin, 2005)
11. L.B. Da Silva, P. Celliers, G.W. Collins, K.S. Budil, N.C. Holmes, T.W. Barbee Jr., B.A. Hammel, J.D. Kilkenny, R.J. Wallace, M. Ross, R. Cauble, A. Ng, G. Chiu, Phys. Rev. Lett. **78**, 483 (1997)
12. S.T. Weir, A.C. Mitchell, W.J. Nellis, Phys. Rev. Lett. **76**, 1860 (1996)

13. G.W. Collins, L.B. Da Silva, P. Celliers, D.M. Gold, M.E. Foord, R.J. Wallace, A. Ng, S.V. Weber, K.S. Budil, R. Cauble, Science **281**, 1178 (1998)
14. M. Mochalov, *Conf. Strongly Coupled Coulomb Systems*, Moscow, 2005; V. Fortov, M. Mochalov et al., Phys. Rev. Lett. **99**, 185001 (2007)
15. N. Nettelmann, B. Holst, A. Kietzmann, M. French, R. Redmer, D. Blaschke, Astrophys. J. **683**, 1217 (2008)
16. D.G. Hummer, D. Mihalas, Astrophys. J. **331**, 794 (1988); D. Mihalas, W. Däppen, D.G. Hummer, Astrophys. J. **331**, 815 (1988)
17. S. Atzeni, J. Meyer-ter-Vehn, *The Physics of Inertial Fusion* (Oxford Science Publication, Oxford, 2004)
18. D. Beule, W. Ebeling, A. Förster, H. Juranek, S. Nagel, R. Redmer, G. Röpke, Phys. Rev. B **59**, 14–177 (1999)
19. D. Beule, W. Ebeling, A. Förster, H. Juranek, R. Redmer, G. Röpke, Contrib. Plasma Phys. **39**, 21 (1999)
20. W. Ebeling, H. Hache, H. Juranek, R. Redmer, G. Röpke, Contr. Plasma Phys. **45**, 160 (2005)
21. E. Beth, G.E. Uhlenbeck, Physica **3**, 729 (1936); **4**, 915 (1937)
22. M. Schmidt, G. Röpke, H. Schulz, Ann. Phys. **202**, 57 (1990)
23. G. Röpke, K. Kilimann, D. Kremp, W.D. Kraeft, Phys. Lett. **68A**, 329 (1978)
24. G. Röpke, K. Kilimann, D. Kremp, W.D. Kraeft, R. Zimmermann, phys. stat. sol. B **88**, K59 (1978); R. Zimmermann, K. Kilimann, D. Kremp, W.D. Kraeft, G. Röpke, phys. stat. sol. B **90**, 175 (1978)
25. R. Redmer, Phys. Rep. **282**, 35 (1997)
26. V.E. Fortov, I.T. Yakubov, *Physics of Nonideal Plasmas* (Hemisphere Publications Corporation, New York, 1990)
27. H.C. Graboske, D.J. Harwood, F.J. Roges, Phys. Rev. **186**, 210 (1969)
28. G. Röpke, T. Seifert, H. Stolz, R. Zimmermann, Phys. Stat. Sol. B **100**, 215 (1980); G. Röpke, M. Schmidt, L. Münchow, H. Schulz, Nucl. Phys. **A 399**, 587 (1983); G. Röpke, in *Aggregation Phenomena in Complex Systems*, ed. by J. Schmelzer et al., (Wiley-VCH, Weinheim, New York, 1999)
29. W. Ebeling, K. Kilimann, Z. Naturforschung **44A**, 519 (1989)
30. S. Arndt, W.D. Kraeft, J. Seidel, phys. stat. sol. B **194**, 601 (1996)
31. W. Stolzmann, W. Ebeling, Phys. Lett. A **248**, 242 (1998)
32. A. Bunker, S. Nagel, R. Redmer, G. Röpke, Phys. Rev. B **56**, 3094 (1997); Contrib. Plasma Phys. **37**, 115 (1997)
33. H. Juranek, R. Redmer, J. Chem. Phys. **112**, 3780 (2000)
34. D. Beule, W. Ebeling, A. Förster, H. Juranek, R. Redmer, G. Röpke,Phys. Rev. E **63**, 060202 (2001)
35. N.F. Carnahan, K.E. Starling, J. Chem. Phys. **51**, 635 (1969)
36. W. Ebeling, A. Förster, V.E. Fortov, V.K. Gryaznov, A. Ya. Polishchuk, *Thermophysical Properties of Hot Dense Plasmas* (Teubner Verlag, Stuttgart and Leipzig, 1991)
37. W. Ebeling, H. Hache, M. Spahn, Eur. Phys. D **23**, 265 (2003)
38. D. Beule, W. Ebeling, A. Förster, Physica A **241**, 719 (1997)
39. W.J. Nellis, A.C. Mitchell, M. van Thiel, G.J. Devine, R.J. Trainor, N. Brown, J. Chem. Phys. **79**, 1480 (1983)
40. N.C. Holmes, M. Ross, W.J. Nellis, Phys. Rev. B **52**, 15–835 (1995)
41. M. Ross, Phys. Rev. B **58**, 669 (1998)
42. H. Shimizu, E.M. Brody, H.K. Mao, P.M. Bell, Phys. Rev. Lett. **47**, 128 (1981)

43. T.J. Lenosky, J.D. Kress, L.A. Collins, Phys. Rev. B **56**, 5164 (1997)
44. F.J. Rogers, D.A. Young, Phys. Rev. E **56**, 5876 (1997)
45. Z. Zinamon, Y. Rosenfeld, Phys. Rev. Lett. **81**, 4668 (1998)
46. J.D. Johnson, Phys. Rev. E **59**, 3727 (1999)
47. M. Ross, Phys. Rev. B **54**, 9589 (1996)
48. W.J. Nellis, S.T. Weir, A.C. Mitchell, Phys. Rev. B **59**, 3434 (1999)
49. L.D. Landau, E.M. Lifschitz, *Lehrbuch der theoretischen Physik, Bd. VI: Hydrodynamik* (Akademie Verlag, Berlin, 1991)
50. R. Redmer, G. Röpke, D. Beule, W. Ebeling, Contrib. Plasma Phys. **39**, 25 (1999)

4

Metal–Insulator Transition in Dense Hydrogen

Ronald Redmer and Bastian Holst

Abstract. We review state-of-the-art theoretical approaches to the metal–insulator transition in dense hydrogen by comparing advanced chemical models with ab initio simulation techniques as well as shock-wave experiments. Chemical models rely on the effective interaction potentials between the different species and a proper calculation of the density- and temperature-dependent partition functions. A common feature of chemical models is the occurrence of a first-order phase transition at high pressures, the plasma phase transition. Ab initio simulation techniques which avoid a discrimination of electron states into *bound* and *free* states by starting from a strict physical picture show up to now no clear signal of a first-order phase transition. However, the metal–insulator transition as experimentally deduced from electrical conductivity and the reflectivity measurements is very well reproduced.

4.1 Introduction

The transition from nonmetallic to metallic behavior has been studied in various systems based on the pioneering ideas of Mott. This book aims to review this effect that occurs in Coulomb systems such as fluids, plasmas, and clusters up to nuclear matter and the quark–gluon plasma. The most prominent example for such a transition is of course the simplest and the most abundant element in nature – hydrogen. Metallization of solid hydrogen at high pressures has been predicted by Wigner and Huntington [1] already in 1935 but not been verified yet. Numerous experimental and theoretical studies were performed since then to reveal details of this electronic transition from solid and liquid hydrogen to a conducting plasma state, see, for example, [2–4].

Experimental investigations of hydrogen at high pressures of several megabar (1 Mbar = 100 GPa) can be performed with diamond anvil cells [5] or dynamically by using strong shock waves generated by high explosives [6–8], gas guns [9–11], high-power lasers [12–14], or magnetically launched flyer plates [15, 16]. The respective data for the equation of state (EOS), sound velocity, electrical conductivity, and reflectivity along the single shock Hugoniot curve or the isentrope in case of multiple shocks clearly indicate a transition from a nonconducting, molecular liquid at low temperatures and pressures through a warm fluid at moderate temperatures of few electron volt and pressures of about 1 Mbar to a conducting fully ionized plasma at still higher temperatures and pressures. The parameters of the transition region characterize warm dense matter (WDM) where strong correlations and quantum effects are important.

Some aspects of this nonmetal-to-metal transition are not clear yet. For instance, chemical models that are reviewed in Sect. 4.3 usually predict that this electronic transition is accompanied by a first-order phase transition with a discrete jump in density and a corresponding heat of transition. Following the ideas of Landau and Zeldovich [17] for the high-pressure phase diagram of matter, the coexistence line between the nonconducting and conducting phase of that plasma phase transition (PPT) including possible triple points and a second critical point has been calculated in numerous chemical models, see, for example [18–26]; see also Chap. 3.

On the other hand, ab initio path integral Monte Carlo (PIMC) and quantum molecular dynamics (QMD) simulations usually indicate a continuous transition without a thermodynamic instability [27, 28], see Sect. 4.4.1 for details. These methods are based on a strict physical picture and avoid the definition of effective density- and temperature-dependent two-particle potentials and cross sections such as the Debye or the polarization potential, see [29]. However, because the particle number and simulation time is limited in these ab initio approaches, we cannot exclude the existence of a PPT yet. This discussion is even further enhanced as first experimental signatures of a PPT were reported recently for deuterium [8] and quantum molecular dynamics simulations showed that a liquid–liquid first-order phase transition can be identified just in the region of the nonmetal-to-metal transition [30–32].

We aim to review here the nonmetal-to-metal transition in dense hydrogen and deuterium based on new results for the EOS, the Hugoniot curve, as well as the electrical conductivity and reflectivity, which are derived from chemical models and large scale QMD simulations. The PPT is discussed in detail. These new results have also a great impact on the state-of-the-art models for planetary interiors for which the thermophysical properties of WDM in general and of hydrogen–helium mixtures in particular are the basic input [33–37].

4.2 Mott Effect in Dense Plasmas

4.2.1 Theoretical Concept

The description of the nonmetal-to-metal transition in dense plasmas was initially based on the chemical picture that considers bound states such as atoms H and molecules H_2 out of the elementary particle electrons e and protons p as new entities. Chemical models consider strong and long-living correlations in form of bound states correctly, but require effective pair potentials for the interaction of the new species with all other particles to describe also their interactions in a dense, partially ionized plasma consistently. Chemical models are well founded as long as the bound states can clearly be defined, which is usually the case at low to moderate densities. The thermophysical properties of dense, partially ionized plasmas were elaborated successfully by calculating the nonideality corrections to the EOS and the laws of mass action as well as to the transport coefficients; see [29].

At high densities, the bound electrons are delocalized due to pressure ionization (Mott effect), and a corresponding nonmetal-to-metal transition occurs. Therefore, chemical models become highly questionable just in the region of the nonmetal-to-metal transition because bound states, no longer exist from a chemical perspective. However, the particles are strongly correlated in that region so that instead of long-living bound states short-living, transient two-particle states or even higher clusters may occur. A complex interplay between screening effects, correlations, and disorder is expected in the transition region so that combined Mott–Hubbard–Anderson models have been developed for that purpose [38, 39]. Chemical models interpolate qualitatively between the limiting cases of a neutral gas or fluid at lower densities with almost no ionization and a fully ionized plasma at high densities.

The first study of the Mott effect in dense plasmas within a consistent physical picture was performed by Rogers, Graboske, and Harwood [40] by solving the Schrödinger equation for the Debye potential numerically. They found that at a critical (Mott) density $\kappa_{\mathrm{Mott}} a_{\mathrm{B}} = 1.19$ all bound states including the 1s ground state disappear; $\kappa = 1/R_{\mathrm{D}}$ is the inverse Debye screening length. Zimmermann et al. [41] treated many-particle effects such as dynamical screening, self-energy, Pauli blocking, and exchange interactions consistently within a Greens function approach by deriving and solving an effective wave or Bethe–Salpeter equation,

$$(H^{0}_{\mathrm{ep}} - z)\Psi_{ab} = H^{\mathrm{pl}}_{ab}(z)\Psi_{ab}. \tag{4.1}$$

The two-particle Hamilton operator for the unperturbed atom is denoted by H^{0}_{ab}, and the many-particle effects mentioned above are contained in a medium or plasma contribution H^{pl}_{ab}. It was shown that the binding energies of the localized electrons in the medium remain almost constant due to strong cancelation effects in H^{pl}_{ab}, and that the energy of the free continuum states is lowered with the density according to $\sim \mathrm{n}^{1/2}$. These pioneering studies provided the theoretical foundation for the much older, intuitive concept of lowering of the ionization energy in dense plasmas, which is illustrated in Fig. 4.1. Further work in this context is reviewed in [42].

However, strong correlations and quantum effects in WDM, especially in the region of the Mott transition, cannot be treated adequately within perturbation theory so that more appropriate methods such as density functional theory (DFT) have to be applied, see Sect. 4.4.1. In addition, one expects that correlations and thermal effects broaden the bound state energies as well as

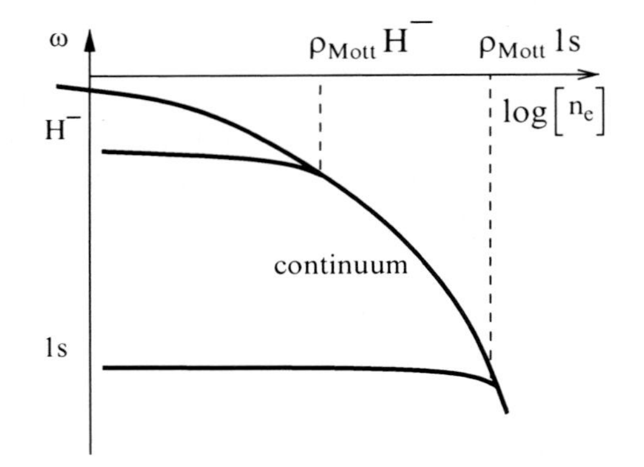

Fig. 4.1. Schematic illustration of pressure ionization (Mott effect) in dense hydrogen plasmas. All binding energies vanish at a specific Mott density. Shown are the hydrogen negative ion H^- and the 1*s* ground state

the continuum edge so that no sharp transition will occur at finite temperatures. This may not prevent the system from going through a thermodynamic instability due to pressure ionization (Mott transition) as will be discussed later

4.2.2 Experimental Signatures

The drastic increase of the ionization degree along the Mott transition leads to strong changes of the thermophysical properties of the plasma. Besides the possible instability in the EOS, which will be discussed later, the electrical conductivity is affected strongest by this ionization transition. While for low and high densities the Spitzer and Ziman formula apply, a minimum in the electrical conductivity is expected for low temperatures at intermediate densities, see [29, 42]. This behavior is illustrated in Fig. 4.2.

The Mott transition region is strongly correlated and can be accessed experimentally only by means of novel techniques. Iermohin et al. [43] were the first to show that a minimum in the electrical conductivity exists in dense metallic vapors, which were compressed adiabatically in shock tubes. Single and multiple shock wave experiments were performed later for the measurement of the electrical conductivity in warm dense hydrogen and deuterium [6, 11]. The Mott transition has now been identified clearly at about 0.6–0.7 g cm^{-3} in hydrogen; for deuterium a factor of two applies due to mass scaling. The transition from a nonconducting molecular fluid to a conducting plasma has also been derived from reflectivity measurements along the single-shock Hugoniot curve [44].

We show in the next sections that a correct description of the nonmetal-to-metal transition in hydrogen and deuterium is only possible within a strict physical picture. Chemical models may give reasonable results but rely

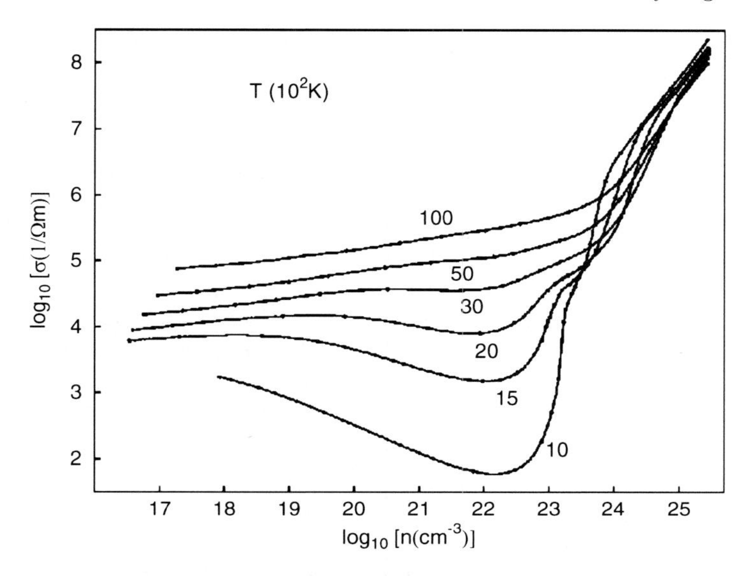

Fig. 4.2. Isotherms for the electrical conductivity as function of the density in hydrogen plasma as derived from a chemical model [23]. While the Spitzer theory applies for low densities, the Ziman formula is valid for high-density degenerate plasmas. The minimum occurs due to partial ionization and is most pronounced at low temperatures. The drastic increase describes the Mott transition

strongly on the treatment of the nonidelity corrections. Therefore, the development and application of ab initio approaches is inevitable to get more insight into the complex physics along the Mott transition.

4.3 Advanced Chemical Models

4.3.1 Free Energy Model for the EOS of Dense Hydrogen

Considering warm dense hydrogen as a partially ionized plasma (PIP) in the chemical picture, a mixture of a neutral component (atoms and molecules) and a plasma component (electrons and protons) is in chemical equilibrium with respect to dissociation and ionization. From an expression for the free energy $F(T,V,N) = F_0 + F_\pm + F_{\mathrm{pol}}$ of the neutral (F_0) and charged particles ($F_\pm$), the equation of state (EOS) can be derived, see [45,46]. The first two terms consist of ideal and interaction contributions and can be written as $F_0 = F_0^{\mathrm{id}} + F_0^{\mathrm{int}}$, and $F_\pm = F_\pm^{\mathrm{id}} + F_\pm^{\mathrm{int}}$. F_{pol} contains interaction terms between charged and neutral components caused by polarization [47].

For the neutral subsystem, the EOS is determined within fluid variational theory (FVT) by calculating the free energy $F_0^{\mathrm{int}}(T,V,N)$ via the

Gibbs–Bogolyubov inequality [48]. This method has been generalized to two-component systems with a reaction [49–51] so that also molecular systems at high pressure can be treated where pressure dissociation occurs, for example, $H_2 \rightleftharpoons 2H$ for hydrogen. In chemical equilibrium, $\mu_{H_2} = 2\mu_H$ has to be fulfilled, and the number of atoms and molecules can be determined self-consistently via the chemical potentials $\mu_c = (\partial F/\partial N_c)_T$. The effective interactions between the neutral species are modeled by exp-6 potentials, and the free energy of a multicomponent reference system of hard spheres has to be known; for details, see [49, 50, 52].

Efficient Padé approximations for the free energy developed by Chabrier and Potekhin [53] were used to treat the charged component; see also [18,54]. The neutral component is in ionization equilibrium $H \rightleftharpoons e+p$ with the charged component and the degree of ionization is fixed by the relation $\mu_H = \mu_e + \mu_p$.

4.3.2 Reduced Volume Concept

There is another interaction between the charged component and the neutral fluid, because point-like particles cannot penetrate into the volume occupied by atoms and molecules, which have a finite size. The result is a correction in the description of the ideal gas of the charged component [55, 56] so that the ideal free energy of protons and electrons $F_\pm^{id}$ is finally dependent on the reduced volume $V^* = V \cdot (1 - \eta)$,

$$F_\pm^{id}(T, V^*, N) = N_\pm k_B T \cdot f_\pm^{id,*}, \tag{4.2}$$

instead of the volume, where η is the ratio of the volume that cannot be penetrated by point-like particles with respect to the total volume. It depends on hard sphere diameters, which result from FVT calculations self-consistently. The free energy density $f_\pm^{id,*}$ was calculated via Fermi integrals so that degeneracy effects are fully taken into account. For an easier handling of the EOS data, for example, as input into planetary models, intersections of the pressure isotherms were avoided by introducing a minimum diameter d_{min}. Starting at low temperatures, it remains almost constant up to 15,000 K, then it increases up to 20,000 K and remains constant again for higher temperatures. Results for the diameter of the hydrogen atom derived from the confined atom model [57] are within the range of this parameter.

The reduced volume concept yields drastic changes of the chemical potential of each component at higher densities, which leads to pressure ionization. This is due to the fact that additional terms appear in the chemical potential via the particle number derivative of the free energy, and thermodynamic functions of degenerate plasmas are very sensitive to these changes in density.

The present model FVT^+ is a generalization of earlier work [52], where only ideal plasma contributions have been treated and includes all interaction contributions to the chemical potentials in an advanced chemical model.

4.3.3 Results for the EOS

The composition of hydrogen plasma as derived from FVT$^+$ for two temperatures is shown in Fig. 4.3. Hydrogen is an atomic gas at low temperatures (left) and low densities. Molecules are formed at medium density due to the corresponding mass action law. In the high-density region, pressure dissociation and ionization take place. The nonideality corrections to the free energy force a transition from a molecular fluid to a fully ionized plasma. The formation of molecules is suppressed and pressure ionization becomes the dominating process at higher temperatures (right). Because of thermal ionization, a fully ionized plasma is found at low densities and high temperatures.

To present thermal EOS data, pressure isotherms are shown in Fig. 4.4 over a wide range of temperatures and densities. The system behaves like a neutral fluid at low densities. Nonideality corrections to the free energy of the neutral subsystem lead to a nonlinear behavior of the isotherms between densities of 10^{-3} and $10^{-1}\,\mathrm{g\,cm^{-3}}$. A phase transition occurs at still higher

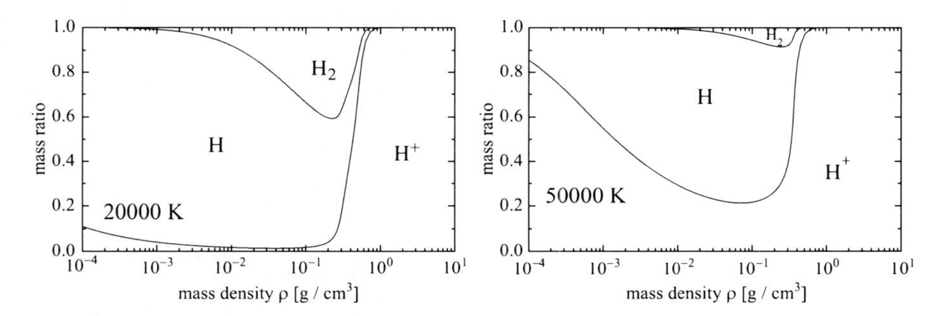

Fig. 4.3. Composition of dense hydrogen plasma for 20,000 K (*left*) and 50,000 K (*right*) [58]

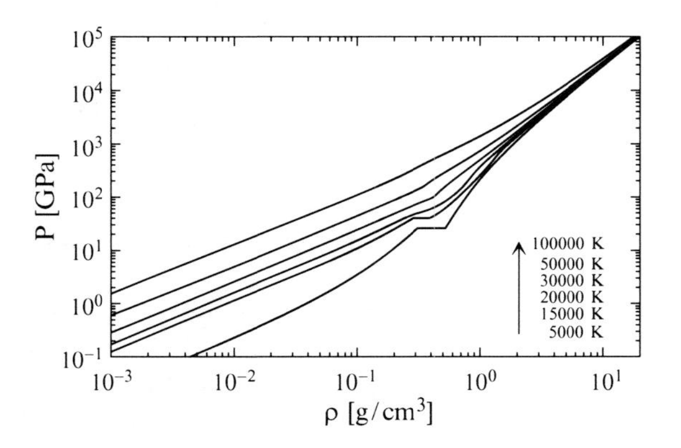

Fig. 4.4. Pressure isotherms for dense hydrogen [58]

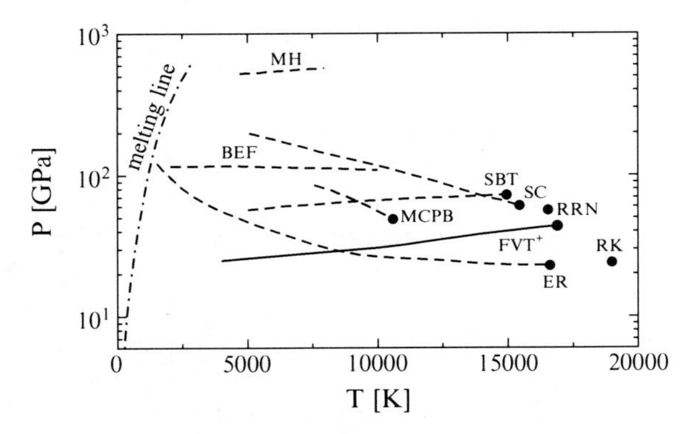

Fig. 4.5. Phase diagram for warm dense hydrogen [58]. The prediction of FVT^+ (*solid line*) is compared with other predictions (*dashed lines*) for the PPT: SC [20,21], RK [59], MH [60], ER [18,54,61], SBT [22], RRN [23], BEF [62], and MCPB [63]

Table 4.1. Theoretical results for the critical point of the hypothetical plasma phase transition (PPT) in hydrogen, which was predicted by Zeldovich and Landau [17] and Norman and Starostin [67]

T_c (10^3 K)	p_c (GPa)	ρ_c ($\mathrm{g\,cm^{-3}}$)	Method	Authors	Reference
12.6	95	0.95	PIP	Ebeling/Sändig (1973)	[64]
19	24	0.14	PIP	Robnik/Kundt (1983)	[59]
16.5	22.8	0.13	PIP	Ebeling/Richert (1985)	[18,54,61]
16.5	95	0.43	PIP	Haronska et al. (1987)	[65]
15	64.6	0.36	PIP	Saumon/Chabrier (1991)	[20]
15.3	61.4	0.35	PIP	Saumon/Chabrier (1992)	[21]
14.9	72.3	0.29	PIP	Schlanges et al. (1995)	[22]
16.5	57	0.42	PIP	Reinholz et al. (1995)	[23]
11	55	0.25	PIMC	Magro et al. (1996)	[63]
20.9	0.3	0.002		Kitamura/Ichimaru (1998)	[66]
16.8	45	0.35	PIP	FVT^+: Holst et al. (2007)	[58]

densities which is treated by a Maxwell construction. The thermodynamic instability vanishes with increasing temperatures, which fixes the critical point at 16,800 K, 0, 35 $\mathrm{g\,cm^{-3}}$, and 45 GPa.

In Fig. 4.5, the critical points and the related coexistence lines as derived from FVT^+ and from other chemical models are compared. The critical point of FVT^+ lies within the range of most of the other predictions, which cover temperatures between 15,000 and 20,000 K. The resulting coexistence line of FVT^+ is, however, lower than most of the other results. The critical parameters of various chemical models are summarized in Table 4.1.

New shock-wave experiments [8] imply that a PPT occurs in deuterium at a density of about 1.5 g cm^{-3} and at a coexistence pressure of about 1 Mbar. Each of these values is twice as high as derived from FVT$^+$ and other chemical models. Therefore, a more detailed analysis of this experiment and the chemical models is needed.

4.4 Warm Dense Hydrogen in the Physical Picture

4.4.1 Quantum Molecular Dynamics Simulations

An efficient tool to describe warm dense matter within a strict physical picture are QMD simulations [28,30,68–76]. QMD simulations combine classical molecular dynamics for the (heavy) ions with a quantum treatment for the electrons within DFT, see [77]. This allows for a systematic treatment of correlation and quantum effects. An alternative approach in this context is wave packet molecular dynamics (WPMD) simulations, in which the electrons are represented on a semiquantal level by wave packets [32,78–82].

Besides the EOS data, a broad spectrum of physical properties of warm dense hydrogen can be determined by QMD simulations. The electronic structure calculations within DFT yield the charge density distribution in the simulation box at every time step, and the molecular dynamics run gives valuable structural information via the ion–ion pair correlation function. This is important for the identification and characterization of phase transitions such as solid–liquid or liquid–plasma as well as for the nonmetal-to-metal transition.

The basis of DFT is given by the theorems of Hohenberg and Kohn [83] and provides the electron density that minimizes the ground state energy of the system. It has been proven that this density is a unique functional of the effective potential $V_{\rm eff}$. To allow practical benefit, Kohn and Sham [84] derived a computational scheme within this formalism, which solves the problem for a fictious system of noninteracting particles that leads to the same electron density. This scheme consists basically of solving the Kohn–Sham equations

$$\left[-\tfrac{\hbar^2}{2m}\nabla^2 + V_{\rm eff}(r)\right]\varphi_k(r) = \epsilon_k\varphi_k(r), \tag{4.3}$$

$$V_{\rm eff}[\varrho(\mathbf{r})] = \int \tfrac{\varrho(\mathbf{r}')e^2}{|\mathbf{r}-\mathbf{r}'|}\,\mathrm{d}\mathbf{r}' - \sum_{k=1}^{N}\tfrac{Z_k e^2}{|\mathbf{r}-\mathbf{R}_k|} + V_{XC}[\varrho(\mathbf{r})].$$

The ab initio QMD simulations [28] were done within Mermin's finite temperature density functional theory (FT-DFT) [85], which is implemented in the plane wave density functional code VASP (Vienna Ab Initio Simulation Package) [86–88]. The projector augmented wave potentials [89] were used and the generalized gradient approximation (GGA) using the parameterization of PBE [90] was applied. Extensive test calculations were performed and have shown that the EOS data are dependent on the plane wave cutoff. A convergence of better than 1% is secured for $E_{\rm cut} = 1{,}200\,\mathrm{eV}$, which is in agreement with the results found already by Desjarlais [72]. This plane wave cutoff was used in all actual calculations. The electronic structure calculations were performed for a given array of ion positions, which are subsequently varied by the forces obtained within the DFT calculations via the Hellmann–Feynman theorem for each molecular dynamics step. This procedure is repeated until the EOS measures are converged and thermodynamic equilibrium is reached.

In a supercell with periodic boundary conditions, the simulations were done for 64 atoms. The temperature of the ions was controlled by a Nosé thermostat [91] and the temperature of the electrons was fixed by Fermi weighting the occupation of bands [87]. The Brillouin zone was sampled by evaluating the results at Baldereschi's mean value point [92], which showed best agreement with a sampling of the Brillouin zone using a higher number of **k**-points. The density of the system was fixed by the size of the simulated supercell. To minimize the statistical error due to fluctuations, the system was simulated 1,000–1,500 steps further after reaching the thermodynamic equilibrium. By averaging over all particles and simulation steps in equilibrium, the EOS data and pair correlation functions were then obtained .

In DFT calculations, the zero-point vibrational energy of the H_2 molecules is not included. The energy $\frac{1}{2}h\nu_{\mathrm{vib}}$ per molecule is added, which is very important, especially at low temperatures and for the calculation of an exact initial internal energy for the reference state of the Hugoniot curve, which is $0.0855\,\mathrm{g\,cm^{-3}}$ at 20 K. To account for this quantum effect adequately for arbitrary temperatures, the fraction of molecules has to be derived, for example, for all states along the Hugoniot curve. This can be done via the coordination number

$$K(r) = \frac{N-1}{V}\int_0^r 4\pi r'^2 g(r')\,\mathrm{d}r', \tag{4.4}$$

which is a weighted integral over the pair correlation function $g(r)$ of the ions. N denotes the number of ions and V the volume of the supercell in the simulation. The doubled value of K at the maximum of the molecular peak in $g(r)$, which is found around $r = 0.748$ Å, is then equal to the fraction of ions bound to a molecule and twice the amount of molecules in the supercell.

For several isotherms, the dissociation degree is calculated and the results are approximated by Fermi functions with two adjustable parameters. The parameters are represented by temperature-dependent functions so that the dissociation degree and, subsequently, the contribution of molecules to the zero-point internal energy are determined for arbitrary temperatures. The results show that above 10,000 K molecules can be neglected.

The resulting dissociation degree obtained by this method [28] is compared with that given by Vorberger et al. [68] in Fig. 4.6 who counted all pairs of atoms in a range of $1.8a_{\mathrm{B}}$ as atoms. Alternatively, the number of molecules was reduced by counting only those pairs that are stable for longer than ten vibrational periods. In all three cases, the amount of molecules is lower for higher densities and the molecules disappear at higher temperatures due to

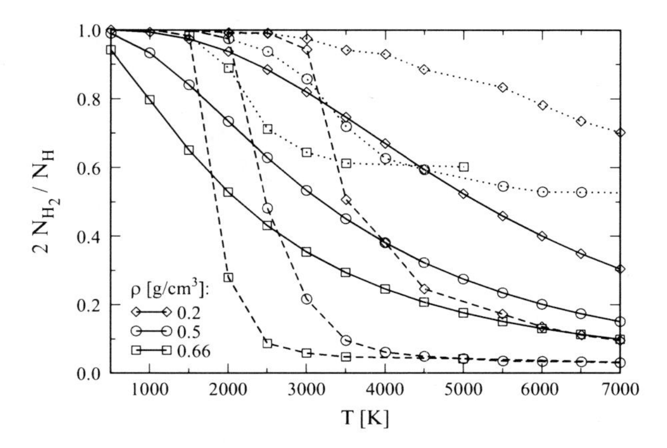

Fig. 4.6. Ratio of hydrogen molecules with respect to the total number of protons for three densities [28]. Our coordination number method (*solid*) is compared with the pair-counting method of Vorberger et al. [68] (*dotted*). Their result counting only pairs with a lifetime longer than ten H_2 vibrational periods is also given (*dashed line*)

thermal dissociation. This picture shows that the dissociation degree depends strongly on the definition of the term *molecule* in the warm dense matter region. The first method described here gives a smoother behavior of the dissociation degree which starts at lower temperatures and is in between the two cases described in [68] at higher temperatures. However, the consequence of these difference is rather small for the EOS data, because in the physical picture the dissociation degree is used only as a factor for the vibrational energy of the molecules, which itself is small compared to the internal energy obtained by the QMD simulations.

4.4.2 Ab Initio EOS Data and Hugoniot Curve

The thermal EOS of warm dense hydrogen is shown in Fig. 4.7. The isotherms of the pressure show a systematic behavior in terms of the density and temperature. In contrast to the findings in the chemical picture, no instability along the pressure isotherms $(\partial P/\partial V)_T > 0$ is found, which would indicate a first-order plasma phase transition (PPT).

Furthermore, the QMD data are, in Fig. 4.7, compared with the chemical models FVT [58] and SCvH-i [33]. The EOS derived by Saumon et al. [33] shows also a PPT (SCvH-ppt data set). The modified SCvH-i data set shown here avoids the PPT by using an interpolation through the instability region. Therefore, both data sets can be used to study the influence of a PPT on the interior models of giant planets such as Jupiter. Consistent chemical models yield the correct low-temperature and low-density limit and agree with our QMD results there. A good agreement is also found in the high-density limit where a nearly temperature-independent behavior characteristic

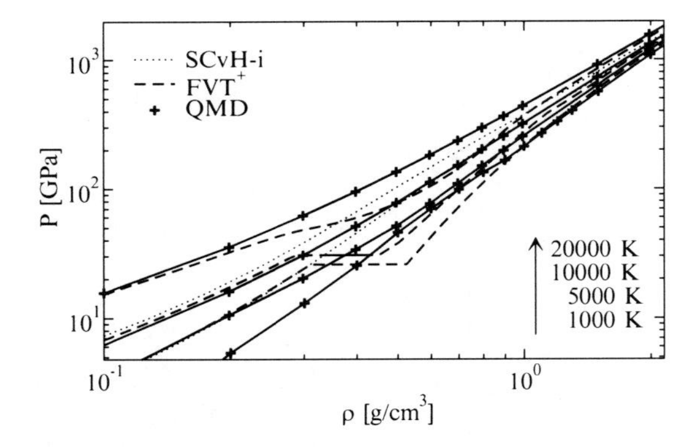

Fig. 4.7. Thermal EOS for warm dense hydrogen (pressure isotherms): QMD data [28] are compared with the chemical models FVT$^+$ [58] and SCvH-i [33]

of a degenerate plasma is found. At medium densities, the pressure isotherms of FVT and SCvH-i lie well below the QMD data; the deviations amount up to 25%.

Within the QMD results, a region with $(\partial P/\partial T)_V < 0$, which has been found previously [30, 68], is encountered. The isochore at $r_\mathrm{s} = 1.75$ given in [68] is reproduced within the uncertainties of the simulations. The instability region appears at pressures up to 200 GPa at temperatures between 1,000 and 4,000 K. It can be related to the rapid dissociation transition at such low temperatures, which leads to a drastic increase of the electrical conductivity, see Sect. 4.4.3. We note that a first experimental signature of an instability has been found in this domain [8], and that recent WPMD simulations [32] show an instability there. The acquisition of still more accurate EOS data for warm dense hydrogen is absolutely essential to solve this challenging problem, see also [31].

A fit the QMD results for the pressure P and the internal energy U by expansions in terms of density ρ and temperature T to allow for a simple processing of this ab initio data in hydrodynamic simulations [93] or planetary physics [35]. Such a fit for their respective data was already given by Lenosky et al. [94] and Beule et al. [62]. The pressure is split into an ideal and an interaction contribution:

$$P = P^{\mathrm{id}} + P^{\mathrm{int}} = \frac{\rho k_\mathrm{B} T}{m_H} + P^{\mathrm{int}}(\rho, T). \tag{4.5}$$

The QMD data for the pressure P given in kbar can be interpolated by the following expansion for the interaction contribution:

$$P^{\mathrm{int}}(\rho, T) = (A_1(T) + A_2(T)\rho)^{A_0(T)}, \tag{4.6}$$

$$A_i(T) = a_{i0} \exp\left(-\left(\frac{T - a_{i1}}{a_{i2}}\right)^2\right) + a_{i3} + a_{i4}T. \tag{4.7}$$

Table 4.2. Coefficients a_{ik} in the expansion for the pressure P^{int} according to (4.6) and (4.7)

i	a_{i0}	a_{i1}	a_{i2}	a_{i3}	a_{i4}
0	0.2234	2919.84	3546.67	1.94023	1.11316×10^{-6}
1	14.7586	2117.98	4559.17	−17.9538	4.88041×10^{-4}
2	−33.8469	2693.63	4159.13	70.582	-2.8848×10^{-4}

Table 4.3. Coefficients b_{jk} in the expansion for the specific internal energy u according to (4.8) and (4.9)

j	b_{j0}	b_{j1}	b_{j2}	b_{j3}	b_{j4}
0	−33.8377	2154.38	3696.89	−300.446	1.77956×10^{-2}
1	55.8794	3174.39	2571.21	56.222	-3.56234×10^{-3}
2	−30.0376	3174.02	2794.39	87.3659	2.0819×10^{-3}
3	5.57328	3215.51	2377.23	−13.1622	-3.84004×10^{-4}
4	−0.3236	3245.48	2991.45	0.682152	2.19862×10^{-5}

By a similar expansion, the QMD data for the specific internal energy $u = U/m$ given in $\mathrm{kJ\,g^{-1}}$ can be given:

$$u = \sum_{j=0}^{4} B_j(T)\rho^j, \tag{4.8}$$

$$B_j(T) = b_{j0} \exp\left(-\left(\frac{T - b_{j1}}{b_{j2}}\right)^2\right) + b_{j3} + b_{j4}T. \tag{4.9}$$

The expansion coefficients a_{ik} and b_{jk} are given in Tables 4.2 and 4.3, respectively.

The expansions (4.6) and (4.8) are valid within a density range from 0.5 to $5\,\mathrm{g\,cm^{-3}}$ between 500 and 20,000 K and reproduce the original ab initio QMD data within 5% accuracy. The expansions fulfill thermodynamic consistency expressed by the relation

$$P - T\left(\frac{\partial P}{\partial T}\right)_V = -\left(\frac{\partial U}{\partial V}\right)_T \tag{4.10}$$

within 15% accuracy, which is mainly due to the deviations of the fit functions from the QMD data itself.

To compare with shock wave experiments, the principal Hugoniot curve was derived from the QMD EOS data, which is a crucial measure for theoretical EOS data and which is plotted in Fig. 4.8. It connects all possible final states (ρ, P, u) of shock wave experiments according to the Hugoniot equation

$$u - u_0 = \frac{1}{2}(P + P_0)\left(\frac{1}{\rho_0} - \frac{1}{\rho_0}\right) \tag{4.11}$$

starting at the same initial conditions (ρ_0, P_0, u_0). We compare with experiments for deuterium, because most of the recent measurements have focused on this isotope of hydrogen. Deuterium has a liquid density of $0.171\,\mathrm{g\,cm^{-3}}$, which can be scaled by a factor of 0.5 to achieve the corresponding state in the hydrogen EOS. The principal Hugoniot curve starts in the liquid with a density of $\rho_0 = 0.0855\,\mathrm{g\,cm^{-3}}$ at a temperature of 20 K and an internal energy of $u_0 = -314\,\mathrm{kJ\,g^{-1}}$. The initial pressure P_0 can be neglected compared with the high final pressure P.

Gas guns [95], magnetically launched flyer plates at Sandia's Z machine [16], or high explosives (HE) [7] have been used to perform shock wave experiments for deuterium. A maximum compression of 4.25 at about 50 GPa is indicated by these experiments.

Systematic deviations from these experiments show another series of laser-driven experiments [96]. Especially, a maximum compression of 6 has been reported at about 1 Mbar. New laser-driven single shock compression experiments indicate a maximum compression of 5 above 1 Mbar [14]. According to the unanimous evaluation of the shock-wave experimental data for molecular liquids [11], the QMD data are compared in Fig. 4.8 only with the data sets mentioned above. The clarification of the discrepancies between the various experimental techniques remains a subject of further work.

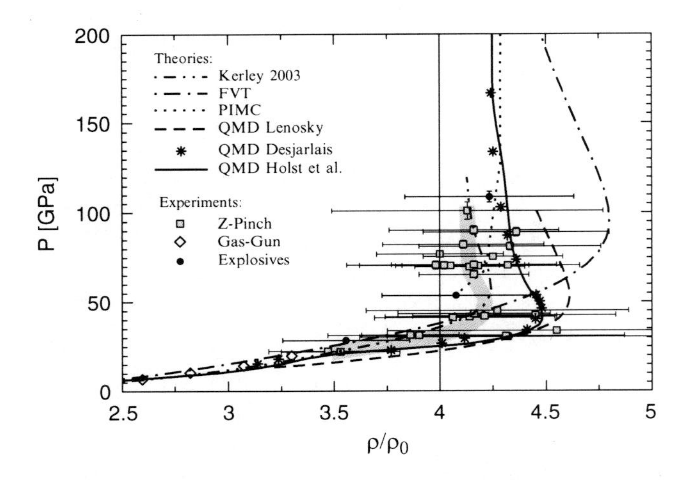

Fig. 4.8. Principal Hugoniot curve for deuterium: results of the QMD simulations [28] (*solid line*) are compared with previous QMD results of Lenosky et al. [97] (*dashed*) and Desjarlais [72] (*stars*), PIMC simulations [100] (*dotted*), the model EOS of Kerley [99] (*dot–dot–dashed*), and the chemical model FVT [49] (*dot–dashed*). Experiments: gas gun [95] (*diamonds*), Sandia Z machine [16] (*grey squares; grey line*: running average through the u_s-u_p data), high explosives [7] (*black circles*)

To achieve fully converged results, a systematic increase of the cutoff energy E_{cut} in QMD simulations from 500 [97] to 1,200 eV[72] was necessary. The converged results are now in agreement with the experimental points. The consideration of the zero-point vibrations of the molecules along the entire Hugoniot curve yields a very good agreement of QMD data with the gas gun experiments [95] especially for lower pressures. The maximum compression of 4.5 of the calculated Hugoniot curve is slightly higher than the HE and Z experiments indicate (about 4.25). Nevertheless, this is an agreement of about 5% accuracy, which can be translated into an accuracy of about 1% in the measured shock and particle velocity. This deviation is in the range of the systematic errors of the experiments. The compression reaches the correct high-temperature limit as given by the PIMC simulations [98] where it decreases with higher pressures and temperatures. For compression rates between 3 and 4, the QMD curve lies slightly below the experimental data, which could be related to the known band gap problem of DFT in GGA. The slope changes slightly due to dissociation of the molecules at a compression ratio of 3.5, which was also observed in similar calculations [30].

As a representative of advanced chemical models that lead to maximum compressions well beyond 4.5 in general, the FVT curve [49] is plotted here. The revised Sesame curve of Kerley [99] agrees with the experiments and shows a maximum compression of 4.25, but the pressure there is slightly higher than the QMD results.

4.4.3 Dynamic Conductivity

The starting point for the evaluation of the dynamic conductivity $\sigma(\omega)$ from which the dielectric function $\varepsilon(\omega)$, the reflectivity, and the dc conductivity can be extracted is the Kubo–Greenwood formula [101, 102]:

$$\sigma(\omega) = \frac{2\pi e^2 \hbar^2}{3m^2\omega\Omega} \sum_{\mathbf{k}} W(\mathbf{k}) \sum_{j=1}^{N} \sum_{i=1}^{N} \sum_{\alpha=1}^{3} [F(\epsilon_{i,\mathbf{k}}) - F(\epsilon_{j,\mathbf{k}})] \times |\langle \Psi_{j,\mathbf{k}} | \nabla_\alpha | \Psi_{i,\mathbf{k}} \rangle|^2 \delta(\epsilon_{j,\mathbf{k}} - \epsilon_{i,\mathbf{k}} - \hbar\omega), \tag{4.12}$$

where e is the electron charge and m its mass. The summations over i and j run over N discrete bands considered in the electronic structure calculation for the cubic supercell volume Ω. The three spatial directions are averaged by the α sum. $F(\epsilon_{i,\mathbf{k}})$ describes the occupation of the ith band corresponding to the energy $\epsilon_{i,\mathbf{k}}$ and the wavefunction $\Psi_{i,\mathbf{k}}$ at $\mathbf{k}$. The δ-function is broadened with a Gaussian because a discrete energy spectrum results from the finite simulation volume [71]. Integration over the Brillouin zone is performed by sampling special $\mathbf{k}$ points [103], where $W(\mathbf{k})$ is the respective weighting factor. To reach a convergence of better than 10% accuracy, Baldereschi's mean value point [92] is used.

The dc conductivity, which follows in the static limit $\omega \to 0$ from the dynamic conductivity $\sigma(\omega)$, is shown in Fig. 4.9. Our calculations along the

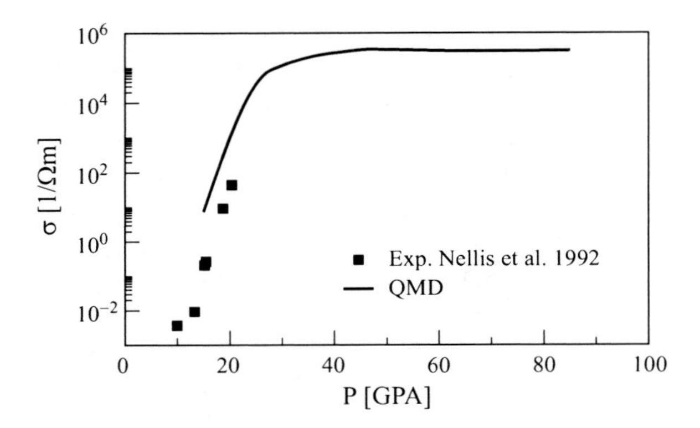

Fig. 4.9. DC conductivity for hydrogen: QMD results along the Hugoniot curve [28] are compared with single shock data of Nellis et al. [104]

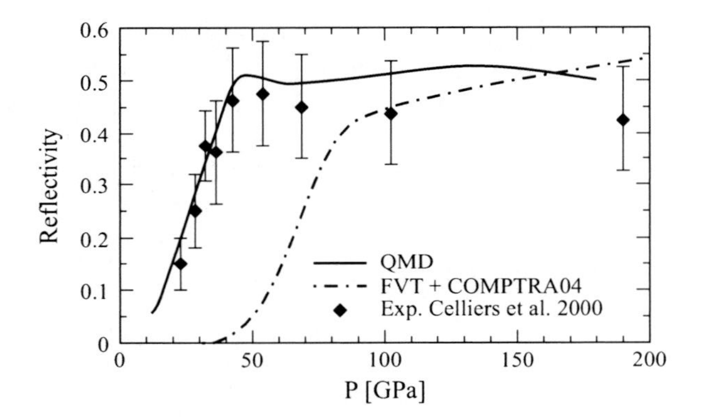

Fig. 4.10. Reflectivity for a wavelength of 808 nm along the Hugoniot curve of hydrogen: QMD results [28] are compared with experimental data of Celliers et al. [44] and predictions of the chemical model FVT [46] using the COMPTRA code [105] and QMD simulations of Collins et al. [70]

principal Hugoniot curve show a rapid increase of the dc conductivity, which is in very good agreement with the single shock experiments of Nellis et al. [104]. At pressures higher than 40 GPa, the QMD results indicate a metallic-like conductivity, which is also apparent in the reflectivity, see Fig. 4.10.

From the frequency-dependent conductivity $\sigma(\omega)$, the optical properties of the liquid can be derived, see (4.12). The imaginary part of $\sigma(\omega)$ is obtained via the Kramers–Kronig relation:

$$\sigma_2(\omega) = -\frac{2}{\pi} \mathrm{P} \int \frac{\sigma_1(\nu)\omega}{(\nu^2 - \omega^2)} \mathrm{d}\nu, \tag{4.13}$$

P is the principal value of the integral. With the complex conductivity, the dielectric function can be calculated directly:

$$\epsilon_1(\omega) = 1 - \frac{1}{\epsilon_0 \omega} \sigma_2(\omega), \tag{4.14}$$

$$\epsilon_2(\omega) = \frac{1}{\epsilon_0 \omega} \sigma_1(\omega). \tag{4.15}$$

The square of the index of refraction, which contains the real part n and the imaginary part k, is equal to the dielectric function. Defined that way, the index of refraction is given by

$$n(\omega) = \sqrt{\frac{1}{2} \left(|\epsilon(\omega)| + |\epsilon_1(\omega)| \right)}, \tag{4.16}$$

$$k(\omega) = \sqrt{\frac{1}{2} \left(|\epsilon(\omega)| - |\epsilon_1(\omega)| \right)}, \tag{4.17}$$

so that the reflectivity r can be calculated via

$$r(\omega) = \frac{[1 - n(\omega)]^2 + k(\omega)^2}{[1 + n(\omega)]^2 + k(\omega)^2}. \tag{4.18}$$

In Fig. 4.10, the QMD results are compared with reflectivities measured along the Hugoniot curve [44]. The agreement is excellent.

The change of the hydrogen reflectivity with pressure can be interpreted as a gradual transition from a molecular insulating fluid through an atomic fluid above 20 GPa, where the atoms have strongly fluctuating bonds with next neighbors [70] to a dense, almost fully ionized plasma with a reflectivity of about 50–60% at high pressures above 40 GPa. The chemical model [46] shows the same behavior, but the abrupt increase of the reflectivity occurs at a higher density. This shows the difficulties of chemical models in finding the correct shifts of the dissociation and ionization energies as function of density and temperature and, thus, the location of the nonmetal-to-metal transition.

4.5 Conclusion

We have reviewed experimental results and theoretical approaches for the nonmetal-to-metal transition in warm dense hydrogen. *Advanced chemical models* such as FVT$^+$ or SCvH are able to describe the limiting cases, the nonconducting dense fluid, and the fully ionized hot plasma, but have conceptual difficulties to describe the transition region where the binding energies of atoms and molecules vanish, and a clear discrimination between bound and free electrons is not possible, see Sect. 4.3. In this region, correlation and quantum effects have to be treated consistently in the EOS and the respective mass action laws by applying, for example, perturbation theory [29] or fluid variational theory [50]. Measured physical quantities such as the Hugoniot curve [45,52,58] and the electrical conductivity [23,105] can be reproduced at least qualitatively. The most striking feature of almost all chemical models is the occurrence of a first-order phase transition as a consequence of the drastic changes of the electronic properties in the transition region. The location of the coexistence line and of a second critical point is sensitive with respect to the approximations made for the nonideality corrections to the equation of state. Alternatively, QMD simulations have been performed within a strict *physical picture*. The resulting thermophysical quantities such as the EOS and Hugoniot curve, reflectivity and conductivity are in very good agreement with shock wave experiments [28], so that a consistent description of warm dense hydrogen can now be given. Although no clear signs of a first-order phase transition have been found so far, this issue remains a subject of ongoing research. Especially, a region with $(\partial P/\partial T)_V < 0$ has been found previously in QMD simulations [28,30,68] at pressures up to 200 GPa and temperatures between 1,000 and 4,000 K. It is related to the rapid, pressure-driven dissociation transition in the low-temperature fluid, which leads to a drastic increase of the electrical conductivity, see Sect. 4.4.3. Note that first experimental signatures of an instability have been found just in this domain [8], and that recent WPMD simulations [32] show a pronounced instability there. Therefore, the acquisition of still more accurate EOS data for warm dense hydrogen is an essential prerequisite to reveal more details of the high-pressure phase diagram of the simplest element in nature. This would also lead to improved models for the interiors of solar and extrasolar planets, see [35].

Acknowledgment

We acknowledge helpful discussions with Michael P. Desjarlais, Werner Ebeling, Martin French, Vladimir E. Fortov, Victor K. Gryaznov, Friedrich Hensel, Wolf Dietrich Kraeft, André Kietzmann, Winfried Lorenzen, Thomas R. Mattsson, Victor B. Mintsev, Nadine Nettelmann, Heidi Reinholz, Gerd Röpke, Christian Toepffer, and Günter Zwicknagel. This work was supported by the Deutsche Forschungsgemeinschaft within the SFB 652 *Strong Correlations and Collective Phenomena in Radiation Fields.*

References

1. E.P. Wigner, H.B. Huntington, J. Chem. Phys. **3**, 764 (1935)
2. M. Städele, R.M. Martin, Phys. Rev. Lett. **84**(26), 6070 (2000)
3. K. Nagao, S.A. Bonev, N.W. Ashcroft, Phys. Rev. B **64**(22), 224111 (2001)
4. W. Ebeling, G. Norman, J. Stat. Phys. **110**(3–6), 861 (2003)
5. H.K. Mao, R.J. Hemley, Rev. Mod. Phys. **66**(2), 671 (1994)
6. V.E. Fortov, V.Y. Ternovoi, M.V. Zhernokletov, M.A. Mochalov, A.L. Mikhailov, A.S. Filimonov, A.A. Pyalling, V.B. Mintsev, V.K. Gryaznov, I.L. Iosilevskii, J. Exp. Theor. Phys. **97**, 259 (2003)

7. G.V. Boriskov, A.I. Bykov, R.I. Il'Kaev, V.D. Selemir, G.V. Simakov, R.F. Trunin, V.D. Urlin, A.N. Shuikin, W.J. Nellis, Phys. Rev. B **71**, 092104 (2005)
8. V.E. Fortov, R.I. Ilkaev, V. Arinin, V.V. Burtzev, V.A. Golubev, I.L. Iosilevkiy, V.V. Khrustalev, A.L. Mikhailov, M.A. Mochalov, V.Y. Ternovoi, M.V. Zhernokletov, Phys. Rev. Lett. **99**, 185001 (2007)
9. S.T. Weir, A.C. Mitchell, W.J. Nellis, Phys. Rev. Lett. **76**, 1860 (1996)
10. W.J. Nellis, S.T. Weir, A.C. Mitchell, Phys. Rev. B **59**, 3434 (1999)
11. W.J. Nellis, Rep. Prog. Phys. **69**, 1479 (2006)
12. L.B. Da Silva, P. Celliers, G.W. Collins, K.S. Budil, N.C. Holmes, T.W. Barbee, B.A. Hammel, J.D. Kilkenny, R.J. Wallace, M. Ross, R. Cauble, A. Ng, G. Chiu, Phys. Rev. Lett. **78**(3), 483 (1997)
13. G.W. Collins, L.B. Da Silva, P. Celliers, D.M. Gold, M.E. Foord, R.J. Wallace, A. Ng, S.V. Weber, K.S. Budil, R. Cauble, Science **281**(5380), 1178 (1998)
14. D.G. Hicks, T.R. Boehly, P.M. Celliers, J.H. Eggert, S.J. Moon, D.D. Meyerhofer, G.W. Collins, Phys. Rev. B **79**, 014112 (2009)
15. M.D. Knudson, D.L. Hanson, J.E. Bailey, C.A. Hall, J.R. Asay, W.W. Anderson, Phys. Rev. Lett. **87**, 225501 (2001)
16. M.D. Knudson, D.L. Hanson, J.E. Bailey, C.A. Hall, J.R. Asay, C. Deeney, Phys. Rev. B **69**, 144209 (2004)
17. Y.B. Zeldovich, L.D. Landau, Zh. Eksp. Teor. Fiz. **14**, 32 (1944)
18. W. Ebeling, W. Richert, Phys. Lett. A **108**, 80 (1985)
19. D. Saumon, G. Chabrier, Phys. Rev. Lett. **62**(20), 2397 (1989)
20. D. Saumon, G. Chabrier, Phys. Rev. A **44**, 5122 (1991)
21. D. Saumon, G. Chabrier, Phys. Rev. A **46**, 2084 (1992)
22. M. Schlanges, M. Bonitz, A. Tschttschjan, Contrib. Plasma Phys. **35**, 109 (1995)
23. H. Reinholz, R. Redmer, S. Nagel, Phys. Rev. E **52**, 5368 (1995)
24. A.A. Likalter, Phys. Rev. B **53**(8), 4386 (1996)
25. A.A. Likalter, J. Exp. Theor. Phys. **86**(3), 598 (1998)
26. W. Ebeling, A. Förster, H. Hess, M.Y. Romanovsky, Plasma Phys. Contr. Fusion **38**(12A), A31 (1996)
27. V.S. Filinov, M. Bonitz, V.E. Fortov, W. Ebeling, P. Levashov, M. Schlanges, Contrib. Plasma Phys. **44**(5–6), 388 (2004)
28. B. Holst, R. Redmer, M.P. Desjarlais, Phys. Rev. B **77**(18), 184201 (2008)
29. R. Redmer, Phys. Rep. **282**, 35 (1997)
30. S.A. Bonev, B. Militzer, G. Galli, Phys. Rev. B **69**, 014101 (2004)
31. S.A. Bonev, E. Schwegler, T. Ogitsu, G. Galli, Nature **431**, 669 (2004)
32. B. Jakob, P.G. Reinhard, C. Toepffer, G. Zwicknagel, Phys. Rev. E **76**, 036406 (2007)
33. D. Saumon, G. Chabrier, H.M. van Horn, Astrophys. J. Suppl. Ser. **99**, 713 (1995)
34. T. Guillot, Science **286**, 72 (1999)
35. N. Nettelmann, B. Holst, A. Kietzmann, M. French, R. Redmer, D. Blaschke, Astrophys. J. **683**, 1217 (2008)
36. W. Lorenzen, B. Holst, R. Redmer, Phys. Rev. Lett. **102**, 115701 (2009)
37. M. French, T.R. Mattsson, N. Nettelmann, R. Redmer, Phys. Rev. B **79**, 054107 (2009)
38. D.E. Logan, Y.H. Szczech, M.A. Tusch, in *Metal-Insulator Transitions Revisited*, ed. by P.P. Edwards, C.N.R. Rao (Taylor and Francis, London, 1995)

39. P.P. Edwards, R.L. Johnston, C.N.R. Rao, D.P. Tunstall, F. Hensel, Phil. Trans. R. Soc. Lond. A **356**(1735), 5 (1998)
40. F.J. Rogers, H.C. Graboske, D.J. Harwood, Phys. Rev. A **1**(6), 1577 (1970)
41. R. Zimmermann, K. Kilimann, W.D. Kraeft, D. Kremp, G. Röpke, Phys. Stat. Sol. B **90**(1), 175 (1978)
42. D. Kremp, M. Schlanges, W.D. Kraeft, *Quantum Statistics of Nonideal Plasmas* (Springer, Heidelberg, 2005)
43. N.N. Iermohin, B.M. Kovaliov, P.P. Kulik, V.A. Riabii, J. Phys. (Paris) **C1**(5), 39 (1978)
44. P.M. Celliers, G.W. Collins, L.B. DaSilva, D.M. Gold, R. Cauble, R.J. Wallace, M.E. Foord, B.A. Hammel, Phys. Rev. Lett. **84**, 5564 (2000)
45. H. Juranek, N. Nettelmann, S. Kuhlbrodt, V. Schwarz, B. Holst, R. Redmer, Contrib. Plasma Phys. **45**(5–6), 432 (2005)
46. R. Redmer, H. Juranek, N. Nettelmann, B. Holst, in *AIP Conference Proceedings 845: Shock Compression of Condensed Matter – 2005*, ed. by M.D. Furnish, M. Elert, T.P. Russell, C.T. White (AIP, Melville, New York, 2006), pp. 127–130
47. R. Redmer, G. Röpke, Physica A **130**(3), 523 (1985)
48. M. Ross, F.H. Ree, D.A. Young, J. Chem. Phys. **79**(3), 1487 (1983)
49. H. Juranek, R. Redmer, J. Chem. Phys. **112**, 3780 (2000)
50. H. Juranek, R. Redmer, Y. Rosenfeld, J. Chem. Phys. **117**, 1768 (2002)
51. Q.F. Chen, L.C. Cai, Y. Zhang, Y.J. Gu, F.Q. Jing, J. Chem. Phys. **124**(7), 074510 (2006)
52. V. Schwarz, H. Juranek, R. Redmer, Phys. Chem. Chem. Phys. **7**(9), 1990 (2005)
53. G. Chabrier, A.Y. Potekhin, Phys. Rev. E **58**, 4941 (1998)
54. W. Ebeling, W. Richert, Ann. Phys. **39**, 362 (1982)
55. A.G. McLellan, B.J. Alder, J. Chem. Phys. **24**, 115 (1956)
56. T. Kahlbaum, A. Förster, Fluid Phase Equil. **76**, 71 (1992)
57. H.C. Graboske Jr., D.J. Harwood, F.J. Rogers, Phys. Rev. **186**, 210 (1969)
58. B. Holst, N. Nettelmann, R. Redmer, Contrib. Plasma Phys. **47**, 368 (2007)
59. M. Robnik, W. Kundt, Astron. Astrophys. **120**, 227 (1983)
60. M. Marley, W. Hubbard, Icarus **73**, 536 (1988)
61. W. Ebeling, W. Richert, W.D. Kraeft, W. Stolzmann, Phys. Stat. Sol. B **104**, 193 (1981)
62. D. Beule, W. Ebeling, A. Förster, H. Juranek, S. Nagel, R. Redmer, G. Röpke, Phys. Rev. B **59**, 14177 (1999)
63. W.R. Magro, D.M. Ceperley, C. Pierleoni, B. Bernu, Phys. Rev. Lett. **76**, 1240 (1996)
64. W. Ebeling, R. Sändig, Ann. Phys. **28**, 289 (1973)
65. P. Haronska, D. Kremp, M. Schlanges, Wiss. Zeit. Univ. Rostock **36**, 98 (1987)
66. H. Kitamura, S. Ichimaru, J. Phys. Soc. Jpn. **67**, 950 (1998)
67. G.E. Norman, A.N. Starostin, High Temp. **8**, 381 (1970)
68. J. Vorberger, I. Tamblyn, B. Militzer, S.A. Bonev, Phys. Rev. B **75**, 024206 (2007)
69. T.J. Lenosky, J.D. Kress, L.A. Collins, I. Kwon, Phys. Rev. B **55**, 11907 (1997)
70. L.A. Collins, S.R. Bickham, J.D. Kress, S. Mazevet, T.J. Lenosky, N.J. Troullier, W. Windl, Phys. Rev. B **63**, 184110 (2001)

71. M.P. Desjarlais, J.D. Kress, , L.A. Collins, Phys. Rev. E **66**, 025401 (2002)
72. M.P. Desjarlais, Phys. Rev. B **68**, 064204 (2003)
73. S. Mazevet, J.D. Kress, L.A. Collins, P. Blottiau, Phys. Rev. B **67**, 054201 (2003)
74. Y. Laudernet, J. Clérouin, S. Mazevet, Phys. Rev. B **70**, 165108 (2004)
75. S. Mazevet, M.P. Desjarlais, L.A. Collins, J.D. Kress, N.H. Magee, Phys. Rev. E **71**, 016409 (2005)
76. S. Mazevet, F. Lambert, F. Bottin, G. Zérah, J. Clérouin, Phys. Rev. E **75**, 056404 (2007)
77. A.E. Mattsson, P.A. Schultz, M.P. Desjarlais, T.R. Mattsson, K. Leung, Model. Simul. Mater. Sci. Eng. **13**, R1 (2005)
78. D. Klakow, C. Toepffer, P.G. Reinhard, J. Chem. Phys. **101**, 10766 (1994)
79. S. Nagel, R. Redmer, G. Röpke, M. Knaup, C. Toepffer, Phys. Rev. E **57**, 5572 (1998)
80. M. Knaup, P.G. Reinhard, C. Toepffer, Contrib. Plasma Phys. **39**, 57 (1999)
81. M. Knaup, G. Zwicknagel, P.G. Reinhard, C. Toepffer, Nucl. Instr. Meth. A **464**, 267 (2001)
82. M. Knaup, P.G. Reinhard, C. Toepffer, G. Zwicknagel, J. Phys. A Math. Gen. **36**, 6165 (2003)
83. P. Hohenberg, W. Kohn, Phys. Rev. **136**, B864 (1964)
84. W. Kohn, L.J. Sham, Phys. Rev. **140**, A1133 (1965)
85. N.D. Mermin, Phys. Rev. **137**, A1441 (1965)
86. G. Kresse, J. Hafner, Phys. Rev. B **47**, 558 (1993)
87. G. Kresse, J. Hafner, Phys. Rev. B **49**, 14251 (1994)
88. G. Kresse, J. Furthmüller, Phys. Rev. B **54**, 11169 (1996)
89. G. Kresse, D. Joubert, Phys. Rev. B **59**, 1758 (1999)
90. J.P. Perdew, K. Burke, M. Ernzerhof, Phys. Rev. Lett. **77**, 3865 (1996)
91. S. Nosé, J. Chem. Phys. **81**, 511 (1984)
92. A. Baldereschi, Phys. Rev. B **7**, 5212 (1973)
93. N.A. Tahir, D.H.H. Hoffmann, A. Kozyreva, A. Tauschwitz, A. Shutov, J.A. Maruhn, P. Spiller, U. Neuner, J. Jakoby, M. Roth, R. Bock, H. Juranek, R. Redmer, Phys. Rev. E **63**, 016402 (2001)
94. T.J. Lenosky, J.D. Kress, L.A. Collins, Phys. Rev. B **56**, 5164 (1997)
95. W.J. Nellis, A.C. Mitchell, M. van Thiel, G.J. Devine, R.J. Trainor, N. Brown, J. Chem. Phys. **79**, 1480 (1983)
96. R. Cauble, L.B.D. Silva, T.S. Perry, D.R. Bach, K.S. Budil, P. Celliers, G.W. Collins, A. Ng, T.W. Barbee Jr., B.A. Hammel, N.C. Holmes, J.D. Kilkenny, R.J. Wallace, G. Chiu, N.C. Woolsey, Phys. Plasma **4**, 1857 (1997)
97. T.J. Lenosky, S.R. Bickham, J.D. Kress, L.A. Collins, Phys. Rev. B **61**, 1 (2000)
98. B. Militzer, D.M. Ceperley, J.D. Kress, J.D. Johnson, L.A. Collins, S. Mazevet, Phys. Rev. Lett. **87**, 275502 (2001)
99. G.I. Kerley, Equation of states for hydrogen and deuterium. Tech. Rep. SAND2003-3613, Sandia National Laboratories (2003)
100. B. Militzer, D.M. Ceperley, Phys. Rev. Lett. **85**, 1890 (2000)
101. R. Kubo, J. Phys. Soc. Jpn. **12**, 570 (1957)
102. D.A. Greenwood, Proc. Phys. Soc. Lond. **71**, 585 (1958)
103. H.J. Monkhorst, J.D. Pack, Phys. Rev. B **13**, 5188 (1976)

104. W.J. Nellis, A.C. Mitchell, P.C. McCandless, D.J. Erskine, S.T. Weir, Phys. Rev. Lett. **68**, 2937 (1992)
105. S. Kuhlbrodt, B. Holst, R. Redmer, Contrib. Plasma Phys. **45**, 73 (2005); The COMPTRA04 source code and data files can be found at http://www.physik.uni-rostock.de/statphys/pages/comptra

5

Resolving the Ion and Electron Dynamics in Finite Systems Exposed to Intense Optical Laser Fields

J. Tiggesbäumker, T. Fennel, N.X. Truong, and K.-H. Meiwes-Broer

Abstract. Clusters show an enhancement in the absorption when exposed to strong optical laser pulses due to an efficient plasmon mediated energy transfer of radiation into the system. This violent interaction transforms the small particle into a dense and hot nanoplasma state. Electrons are efficiently accelerated in the interaction with the laser field and preferentially emitted in the direction of the laser polarization axis at resonance. Corresponding Vlasov-simulations reveal a surface–plasmon rescattering acceleration mechanism, creating ultrashort electron bursts. On the long time scale, the complexes completely disintegrate and highly charged and energetic atomic ions are emitted. The yield of ionized species can be further enhanced by using pulse shaping in connection with a genetic feedback algorithm.

5.1 Introduction

In the past decade, the dynamical response of finite systems exposed to strong laser pulses has become a topic in cluster and plasma physics as well as in the general field of intense laser–matter interaction. The strong field exposure of particles leads to the formation of a dense and high-temperature plasma ball, where correlation effects dominate when probed at sub-relativistic laser intensities ($\ll 10^{18}\,\mathrm{W\,cm^{-2}}$). The finite size of this so-called nanoplasma introduces several interesting features such as electron collisions with the cluster mean field, whose diameter is of the order of nanometer. Moreover, the electron quiver amplitudes under such laser parameter conditions are comparable to the geometrical extension of the cluster, thus each electron in principle probes the whole plasma complex within one laser oscillation cycle. These properties allow to investigate the strong field laser–matter interaction from a point of view which differs from both atomic physics and solid state plasmas and can thus lead to new insights into the dynamical processes. There is another important feature: In strongly excited particles, one has to distinguish between inner and outer ionization as depicted in Fig. 5.1, middle. Extreme cluster ionization leads to an increase in the mean field potential. Thus, an extra outer ionization barrier establishes, which can reach values of the order

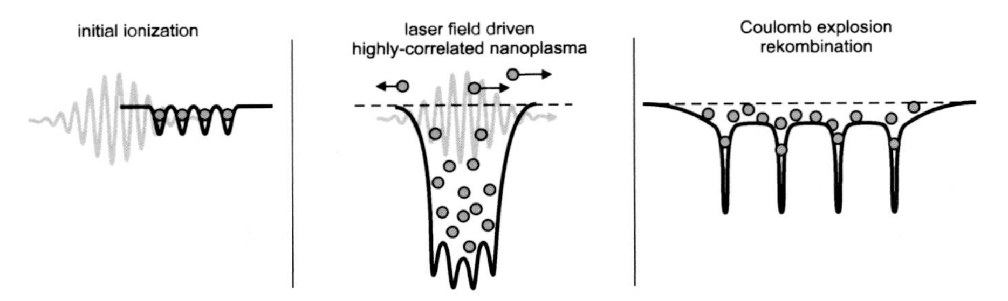

Fig. 5.1. Schematic view of the dynamics that leads to the formation of a nanoplasma and the generation of energetic particles from the interaction of intense laser pulses (10^{13}–10^{16} W cm^{-2}) with clusters. The excitation of rare gas clusters is considered here. *Left*: a partial initial ionization by the leading edge of the pulse leads to an expansion of the cluster; *middle*: the nanoplasma is efficiently heated mostly via resonance absorption. Note, that in contrast to atoms, the cluster has an additional mean-field ionization barrier, which confines the plasma and suppresses outer ionization; *right*: after the laser exposure the complex completely disintegrates by Coulomb explosion, which also includes recombination

of kilo electron-volt. In addition, the strong Coulomb interaction between the partially stripped cores lowers the interatomic ionization barrier. Hence, a broad valence-like band is formed even in rare gas cluster systems. Electron impact excitation of core levels, for example, therefore not necessarily lifts the electron into the continuum but into delocalized states of the mean field potential.

Most of the attention arises from the capability of clusters to absorb a substantial fraction of energy from the laser field. This was to some extent surprising and the results diver vastly from experiments on atoms under comparable atom density conditions. For example, in the early studies at the Imperial College, an extinction coefficient of nearly unity was observed when nanometer-sized rare gas clusters were exposed to 10^{16} W cm^{-2} pulses delivered by an ultrashort laser system [1]. Pioneering work was performed by Rhodes and co-workers who studied the X-ray emission from Xe clusters. Strong evidence was found, that the radiation originates from bound–bound transitions in hollow atoms [2]. The huge energy absorption by the systems is not only reflected in the light emission but also in the generation of energetic particles. Moreover, the cluster completely disintegrates and ions in high charge states are emitted, see Fig. 5.1, right. For example, ion recoil energies as high as 1 MeV [3] and atomic charge states up $q = 28$ have been detected [4,5]. This has led to experiments where nuclear fusion reactions have been studied in high-energy collisions of deuterium atoms from exploding clusters [6].

The Role of Collective Effects

From the theoretical side, the first attempt to describe the cluster response was based on a plasma approach adapted to small particles, called the nanoplasma model [7]. Although the modeling is quite crude, a basic understanding of the interaction could be achieved. Already in this Ansatz, an important parameter was found to be the collective dipole resonance of the delocalized electron gas confined to the cluster volume. As the system size is by far smaller than the typical laser wavelength, only the dipole mode contributes to the absorption. The corresponding resonance frequency, which can be calculated from the ion background density, is well-known and called the Mie or "plasmon" resonance ω_{Mie} [8]. In metallic particles ω_{Mie} is found to be in the ultraviolet. We note, that in intense optical laser systems, the photon energy is fixed to due to technical reasons to a central wavelength of $\lambda = 800\,\text{nm}$ ($\hbar\omega_{\text{L}} = 1.55\,\text{eV}$), that is, a change in the laser frequency to adapt ω_{L} to ω_{Mie} is almost impossible without a substantial loss in pulse intensity.

This leads to a special situation in clusters, bringing the time structure of the ultrashort laser pulse into play, as has been proven in many different experiments, see, for example, [9]. To illustrate how this mismatch could be overcome, we briefly consider silver where the Mie plasmon energy is about 3.5 eV [10]. The value of ω_{Mie} stems from the ion background density n_{bg}. Thus, to drive ω_{Mie} towards ω_{L}, n_{bg} must be lowered, that is, the cluster has to expand. Fortunately, the charging of the cluster by the leading edge of the strong pulse leads to partial ionization of the system via *optical field ionization* [11,12], see Fig. 5.1, left. Simply speaking, the laser field with peak intensity I_0 bends the outer potential barrier, and electrons near the Fermi energy can tunnel through (*tunnel ionization*) or even directly escape from the confining potential (*barrier suppression ionization*, BSI) within half a laser cycle. This turns out to be the main ionization mechanism in an experiment on atoms once with increasing laser field strength a Keldeysh parameter $\gamma = \sqrt{E_{\text{IP}}/2U_{\text{p}}}$ (E_{IP}, electron binding energy; U_{p}, ponderomotive potential) below unity is attained [13]. As a rule of thumb, the value of the ponderomotive potential can be calculated from $U_{\text{p}} = 9.33 \times 10^{-14}\,\text{eV} \times I_0\,[\text{W}\,\text{cm}^{-2}](\lambda[\mu\text{m}])^2$. We emphasize that this process depends only on the pulse intensity I_0 but not on the chosen laser field configuration, for example, the pulse duration.

In clusters, optical field ionization initiates an expansion of the ion structure, leading to a decrease of n_{bg}, thus a red-shift of the resonance towards the laser photon energy $\hbar\omega_{\text{L}}$. Depending on the degree of initial charging, the Coulomb-driven ion–ion separation takes a certain amount of time. For excitations, for example, close to the atomic BSI threshold, the time scales involved are picoseconds, which are large compared to typical laser pulse widths (30–150 fs). Because of the almost complete concentration of oscillator strength in the collective mode, the energy absorption is at a maximum near the Mie resonance. To match both ω_{Mie} and ω_{L}, that is, to profit from the huge absorption capability delivered by the collective mode, it is advantageous to increase the laser pulse duration. By doing this, a substantial fraction of the laser energy can be transferred to the cluster due to the extended interaction time, which brings the system into resonance with the external excitation. In summary, aside the pulse energy, which gives a crude estimate of the possible energy flux into the system, the pulse duration plays an important role in the interaction. An amplitude- and phase-modulated pulse might even be more efficient, and indeed by applying feedback optimized routines to tailor the pulse envelope the system can be driven towards specific configurations, that is, the generation of highly charged ions as first shown in the group of Vrakking [14].

Some Examples

Without going into the details of experiment and simulation, we briefly discuss typical results of our treatments, which test the evolution and contribution of the plasmon in the ionization process. One experimental result is shown in Fig. 5.2. In this study on small gold clusters, the pulse energy (E_L) is constant, whereas the pulse width t is tuned. Consequently, the pulse intensity (E_L/t) decreases for longer pulse durations. In the interaction, the cluster disintegrates completely into atomic ions. On applying a short laser pulse of 140 fs only moderately charged ions ($q \leq 5$) are produced. Adapting the pulse conditions to attain an efficient coupling into the Mie resonance, that is, using a longer pulse duration, the charge state of the ions further increases vastly up to $q = 15$. For even longer pulses ($t \geq 750$ fs), the distribution shifts back to lower charge states. This can partly be attributed to the reduced peak intensity. More precise, the lower initial ionization slows down the expansion, the resonance condition is attained rather in the trailing edge and thus beyond the laser pulse maximum, giving a lower degree of atomic ionization.

Again the key parameters strongly differ from the atomic case, where only the intensity of the pulse drives the strength of the response. Evenmore at comparable pulse intensities, the maximum charge states are found to be substantially higher with respect to atoms. We illustrate the magnitude of the effect. To ionize atomic Xe to q = 20 an intensity of 10^{18} W cm^{-2} is necessary [16], whereas for Xe_N intensities more than three orders of magnitude less are sufficient to generate similar atomic charge states [17], highlighting the huge impact of the collective electron motion on the energy transfer into the cluster. To emphasize the combined action of adapted pulse structure and Mie plasmon absorption in the strong laser field regime, we have termed this mechanism *delayed plasmon enhanced ionization of clusters*.

To check that the plasmon is indeed the main contributor in the ionization of metallic clusters in a strong laser field, Vlasov calculations have been performed using the first and second harmonic (SHG) of the optical laser system, that is, 800 and 400 nm pulses, see Fig. 5.3. In the simulation, two pulses ((800 + 800) nm or (800 + 400) nm) are coupled into the system and the resulting ionization of the *cluster* is evaluated as a function of the time separation Δt of the pulses. The leading laser pulse is chosen to be unchanged and differences in the ionization result only from the impact of the second pulse. Using 400 nm probe pulses, the average charge state decreases with the optical delay. However, with 800 nm pulses, a distinct maximum is found at

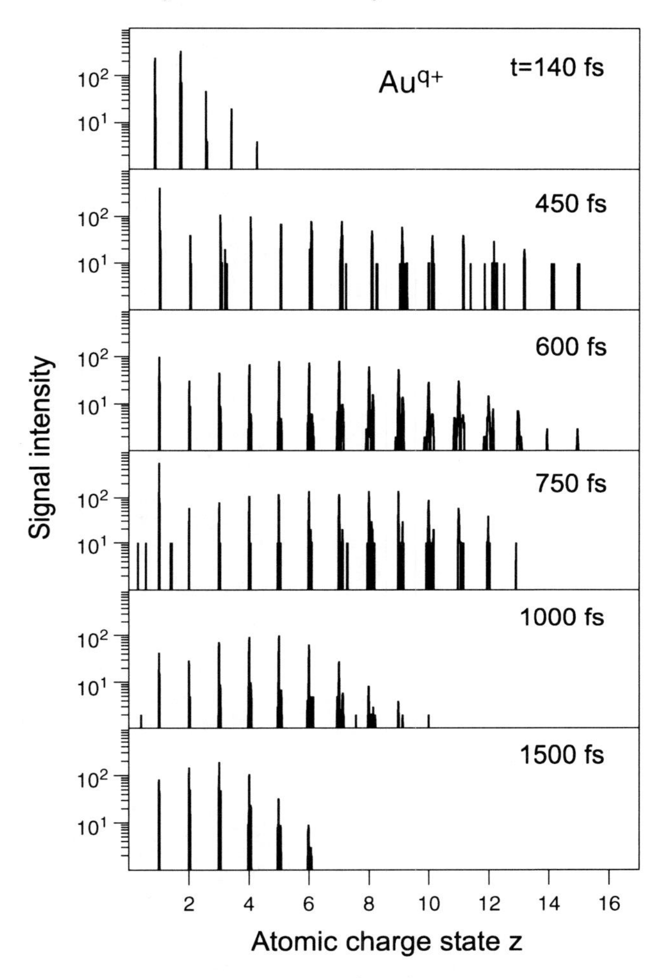

Fig. 5.2. Ionization of small gold clusters (Au_N) irradiated by strong femtosecond laser pulses (6.1×10^{15} W cm^{-2} at 140 fs (FWHM)). For pulse durations of t = 450 fs and 600 fs, the highest atomic charge states are achieved. Adapted from [15]

$\Delta t = 250$ fs. The observation, that a maximum in the dual-pulse delay scan establishes only in the latter scenario is in accordance with an expansion of the cluster, which drives the collective resonance towards *longer wavelengths*. We note that the calculated plasmon energy of the sodium cluster in the initial ground state is 2.84 eV, thus well above the photon energy of 1.55 eV (800 nm) used for excitation. An expansion to a size of approximately two times the initial radius is sufficient to initiate the matching condition between photon energy and plasmon. Thus, the maximum in the dual-pulse delay scan can be associated with plasmon-enhanced ionization, and $\Delta t = 250$ fs reflects the

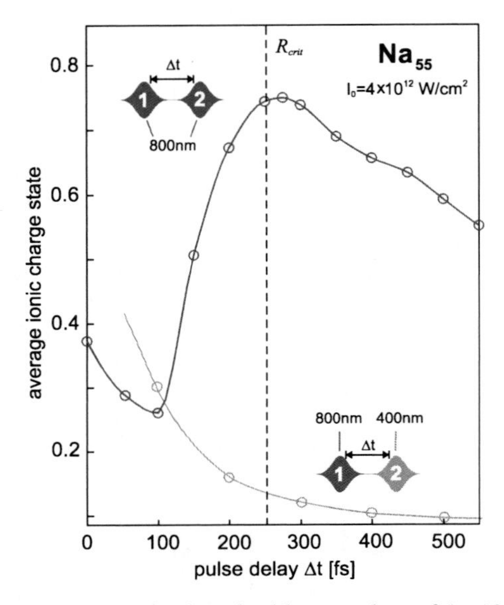

Fig. 5.3. Ionization of Na_{55} exposed to dual laser pulses of 4×10^{12} W cm^{-2} but with different laser wavelengths calculated with the Vlasov code. The simulation shows an increase of the average ionization when only 800 nm pulses are applied, with a maximum at an optical delay of 250 fs. No enhancement is obtained using subsequent 400 nm pulses. The photon energy as well as the delay dependence hint at a leading role of plasmon absorptions in violent cluster excitations. Adapted from [18]

system expansion towards a critical radius, which can be calculated from the Mie formula. For excitation with 400 nm (3.1 eV) postpulses, a blue shift of the resonance towards *shorter wavelengths* is necessary instead. In principle, this can be achieved by a strong initial cluster charging. The pulse intensity of the leading pulse, however, is not sufficient to accomplish this and no maximum establishes. Instead, the cluster expansion drives the plasmon energy of the cluster in a direction away from the laser excitation energy, which is accompanied by a substantial drop in the absorption cross section of the collective mode, reducing the charging efficiency.

In this section, we have demonstrated the contribution of collective effects in the charging of clusters and the possibility to control the energy absorption by adjusting the laser pulse. This can be taken as a starting point to further analyze the ionization dynamics. Briefly, we first concentrate on the experimental procedure (Sect. 5.2) and the computational ansatz (Sect. 5.3) to study the dynamical and nonlinear response of clusters irradiated by pulses from strong ultrashort laser systems. The key results are summarized in Sect. 5.4, giving insight into the ultrafast ionization dynamics. The experimental findings are compared to those obtained from Vlasov molecular dynamics simulations to identify the different mechanisms contributing in the charging process.

5.2 Experimental Challenge

Compared to linear response experiments on clusters (see, for example, the review by deHeer [19]), where large laser beam diameters (several millimeter) can normally be accepted, a tight focusing to some tens of micrometer is necessary to attain intensities of 10^{13}–$10^{16}\,\mathrm{W\,cm^{-2}}$ in the interaction region. The corresponding field strengths are 10^{8}–$10^{11}\,\mathrm{V\,m^{-1}}$. (Note, that these values are on the order of the electric field strength in the hydrogen atom, thus in a regime where the binding conditions are strongly modified by the presence of the external perturbation.) Femtosecond pulses of high energy (several milli-Joules) are thus necessary to allow for a detailed study of the response of particles to strong laser pulse exposure. As an advantage, the ultrashort pulses, in particular the dual-pulse method, enables us to investigate the light-induced dynamics in real time. But the tight focusing of the laser beam brings up a challenge: To have on average at least one particle in the interaction volume, a dense cluster target is necessary, that is, 10^6 clusters per cm^3. Our experimental setup as used in Sect. 5.2 which fulfills this requirement is shown in Fig. 5.4.

The method to form an intense beam of clusters depends on their composition. A high rate of particles can be delivered by a strong supersonic expansion of pressurized and precooled gas atoms. Clusters are formed in the adiabatic cooling process when streaming from the high-pressure region before the nozzle exit into the low-pressure vacuum chamber. Supersonic expansion sources deliver particle densities of up to $10^{15}\,\mathrm{cm^{-3}}$. However, this

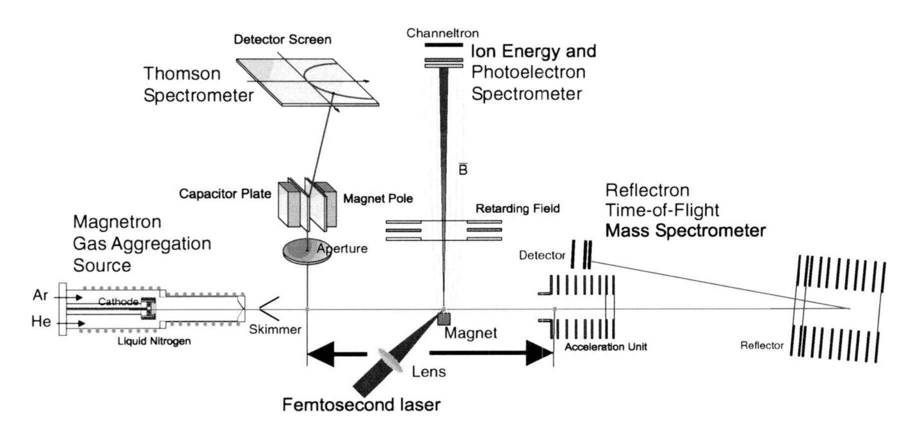

Fig. 5.4. Experimental setup for the investigation of the strong laser pulse interaction with metal clusters. Atoms are brought into the gas phase using a magnetron sputter process. The magnetron is located within an LN2-cooled tube and clusters form by gas aggregation in a 1 mbar mixture of helium and argon buffer gas. The beams expand into vacuum though a nozzle (0.5–3.0 mm). Ion and electron emission resulting from the interaction with the pulse can be analyzed by different time-of-flight methods. In addition, a Thomson spectrometer enables for a simultaneous analysis of energy and momentum of the ionic products

value is achieved only close to the nozzle exit, where the gas load is too high to allow for any operation of a particle detector. In addition, the mean free path might be to small to suppress collisions with the buffer gas completely, which is a strong requirement to extract reliable values like particle energy and emission angle from the experiment. Cluster–cluster interactions might also contribute (cluster–matter effect). Therefore, additional differential pumping stages have to be included downstream the molecular beam axis, reducing the particle density at the point of interaction simply by the geometrical effect of the beam divergence.

Forming Clusters by Gas Aggregation

The beam intensity is further reduced when bare metal clusters are subject of investigation. These are usually formed by either the gas aggregation technique developed by Schulze et al. [20] and extended by the group of Haberland [21] or in a pulsed stream of supersonic helium (Smaller-type source) [22]. In the former types, an oven or a sputtering source is needed whereas in the latter case nanosecond lasers or pulsed discharges [23] are applied to bring the metal partly into the vapor phase. Particle growth proceeds via multiple collisions of metal atoms with the high density buffer gas atoms (He, Ar). Thus, typically in experiments on metal particles, the resulting density is substantially reduced by orders of magnitude when compared to sources producing rare gas clusters. In first studies, we have used the pulsed arc cluster ion source (PACIS) [23] to perform the experiments [9]. In the dual-pulse experiments, the source performance has to be stable over an extended period of time to guarantee good shot–shot target conditions during the optical delay scan. For these studies, we use a home-built Haberland-type of cw magnetron sputter source, which delivers a constant flux of particles, see Fig. 5.4. The cluster size can be varied to a larger extend. Currently, clusters having only a few atoms up to particles of up to 15 nm ($N \approx 10^5$) can routinely be generated with a high flux. Recently, we were able to enhance the performance of this type of source by using aerodynamic focusing, which in principle allows also for a narrowing of the size distribution in the large nanometer regime [24].

Forming Clusters in Helium Nanodroplets

We also take advantage of the pick-up technique to generate particles within a cold and weakly interacting environment. In particular, we use 0.4 K helium nanodroplets to form metal as well as heavier rare gas clusters. This method differs from the single step gas aggregation technique presented above as the clusters are formed in two well-separated steps. First, the droplets are generated in a supersonic expansion of cold helium gas through a 5 μm nozzle into vacuum. By choosing the temperature from 8 to 15 K at the orifice, the average size (log-normal distribution) of the droplets can be varied in between a thousand atoms up to several millions. This offers a broad tunability in the initial size. Typical particle densities are up to 10^{15} droplets per cm^3. Second, the molecular beam transverses a cell containing vapor of the foreign species. Because of the huge geometrical cross-section ($\sigma_{\mathrm{He}} = (2.2\,\mathring{A}) \times N^{1/3}$), atoms from the vapor are caught effectively by the droplets and aggregate within. The energy released in the cluster formation process leads to a slight shrinking of the nanodroplet. As a rule of thumb, about 1,800 helium atoms are evaporated per 1 eV of binding energy released in the aggregation. By varying the density within the pick-up cell, the size distribution of the embedded particles can be chosen almost at will. Using this method, clusters with more than 2,000 atoms can be formed, as shown for magnesium and cadmium [25, 26]. The solid conditions where both initial droplet formation and subsequent foreign particle generation take place and the simplicity in tuning the individual source parameters make the pick-up technique a powerful and versatile method to perform time-resolved studies with high resolution as well as feedback controlled optimization experiments.

Particle Detection

In the interaction, the cluster disintegrates completely when exposed in the central region of the laser focus. However, in the wings, the laser intensity drops below the optical field ionization limit and multiphoton ionization also contributes. Therefore, the ion signal is comprised of charged clusters, fragments, and atoms in high charge states. By selecting ions originating from the laser focus region only, usually no cluster fragments are detected anymore. To determine the recoil energy of the atomic fragments, their time-of-flight towards a detector is measured and thereon converted into kinetic energy. The detection of the fragment mass and the ion charge state is accomplished by mass spectrometry. We use a reflectron time-of-flight setup [27] to identify the products. Figure 5.5 shows a typical charge state spectrum which originates from the interaction of intense laser pulses with cadmium-cluster doped helium droplets, see also Fig. 5.2. As in this contribution we are only interested in the highly charged ion signal, the analysis concentrates on the lower m/q-range. The yield of the Cd^{q+} strongly falls off with increasing charge state. This partly resembles the intensity conditions within the laser focus, favoring the detection of ions generated in the lower intensity wings of the focus, see below. We note that the reflectron mode enables us to focus only a certain ion energy onto the detector. Because of the setup, only the low-recoil energy particles are mapped. More than likely, this underestimates the yield of the higher charge states.

The *electron energies* are verified by time-of-flight measurements as well. For the determination of the integral yield, a magnetic bottle type of instruments [28] is used, whereas in the angular-resolved measurements, the inter-

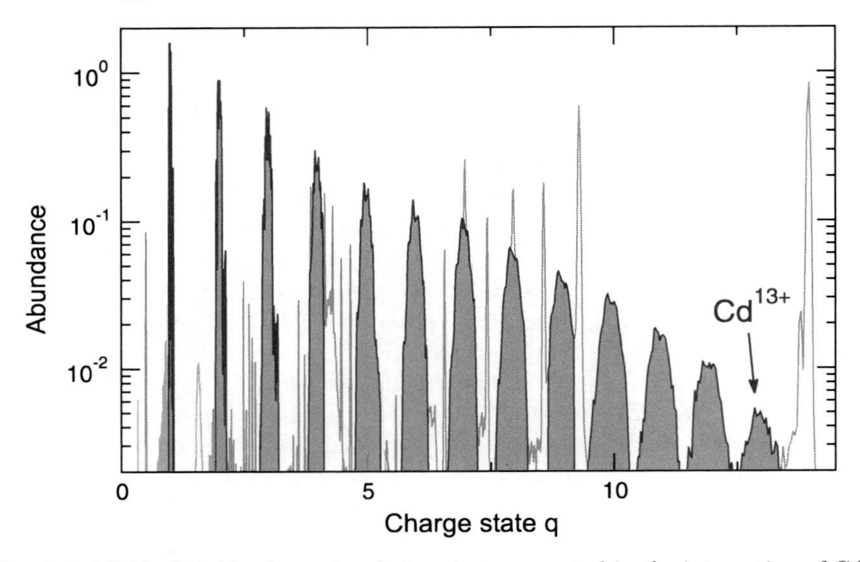

Fig. 5.5. Yield of highly charged cadmium ions generated in the interaction of Cd_N enclosed in helium droplets with 350 fs laser pulses of 3.6×10^{13} W cm^{-2}. Adapted from [18]

action chamber is magnetically isolated by Helmholtz coils and the detection is performed only within a small solid angle. To determine the angular dependence, the laser polarization is rotated.

Ultrafast Laser System

The device to generate pulses in the sub-50 fs regime consists of a Ti:sapphire oscillator (KM Labs, MTS) and a flash lamp driven multipass amplifier (Quantronix, Odin-II HE). The oscillator delivers pulses of 12 fs with a rate of 94 MHz and an energy of 4 nJ. The spectral width of the pulses is about 80 nm (FWHM). These pulses are coupled into the amplifier system. Using the technique of chirped pulse amplification [29], the energy is increased up to 2.5 mJ with a reduced rate of 1 kHz. The initial pulse width is nearly maintained and bandwidth-limited pulses of 35 fs exit the output. Some of the experiments are performed using simple modified pulses. For instance, by detuning the distance of the gratings in the compressor the pulse is stretched. By doing this, pulses having a width of up to several tens of picoseconds can be generated easily. This method was used in the experiment shown in Fig. 5.2.

To map the dynamics in the interaction in more detail but maintaining the pulse intensity, a Mach–Zehnder interferometer is used. This includes a beam attenuator, which is introduced in one of the interferometer arms to generate a pair of pulses of variable optical delay (Δt) and intensity ratio. As in our experiments the subsequent pulse not only probes the dynamics but also dominates the system response, we prefer to call this type of excitation scheme dual-pulse technique instead of the more common term pump–probe, where the second pulse usually introduces only a small perturbation.

In the control studies, we allow the pulse to self-optimize, giving only a certain restriction, for example, generate the highest possible atomic charge state. This attempt is known as feedback-controlled optimization and has led for example, to the concept of coherent control [30, 31]. By introducing a phase shift as well as an amplitude modulation on independent spectral parts of the pulse, a modified pulse that might be more efficient to reach a given goal than the initial pulse leaves the amplifier. The optimization procedure takes advantage of the genetic evolution strategy. Because of the high output energies of the amplifier system in our experiment, the pulse modulation has to be performed between the seeder and the amplifier. Instead of a liquid crystal device as used in many of these type of studies, we use an acousto-optical programmable dispersive filter device (Dazzler, Fastlight) [32]. Collinear phase matching between an RF acoustical and the infrared wave is used to flip the polarization axis of different spectral components at will. Phase shifts of up to 3 ps can be introduced in this way. However, this brings up the influence of nonlinear effects in the amplifier. A one-to-one correspondence between the input pulse from the Dazzler and the pulse coming out of the amplifier is no longer fulfilled. In consequence, assigning a dynamical process in the experiment to a given pulse structure delivered by the Dazzler needs a unique determination of the pulse characteristics entering the interaction volume. In our measurements, we use frequency-resolved optical gating (FROG) [33] to determine the intensity and the phase information of the laser pulse. The result of the pulse analysis that gives a high fitness in the experiment can then be compared to simulations.

5.3 Computational Details

A theoretical description of cluster absorption and ionization in a strong laser field requires suitable methods that allow to resolve the coupled and highly nonadiabatic electron and ion dynamics. As a full quantum mechanical description is not feasible under these circumstances, simplified methods, ranging from molecular dynamics up to quantum approaches based on the time-dependent local-density approximation [34,35], have to be used. A particularly challenging aspect for metal clusters is the presence of delocalized electrons and the fermionic nature of the dynamics. Starting from a fully degenerate state with complete Pauli blocking, the electronic system experiences highly nonequilibrium conditions with significant Coulomb coupling and partial degeneracy within the interaction process. As a result, also electron–electron collisions (EEC), which require a quantum statistical treatment under these conditions, can transiently become very important. Further, the highly nonlinear collective motion of electrons has to be incorporated self-consistently.

Along this line, Vlasov and Vlasov–Uehling–Uhlenbeck (VUU) methods are very appealing for the treatment of metal clusters in strong fields. These semiclassical kinetic methods provide an approximate description of the quantum dynamics without directly referring to wavefunctions [36]. In the following, a concise description of basic methodic aspects and corresponding examples for the energy absorption and ionization of sodium clusters are presented.

Semiclassical Vlasov and VUU Methods

The starting point is the semiclassical approximation of the electron dynamics in the absence of EEC. On the mean-field level, the quantal time-evolution of the electronic system can be described by the von Neumann equation

$$i\hbar\tfrac{\partial}{\partial t}\rho = \left[\tfrac{-\hbar^2}{2m}(\nabla^2_{\mathbf{r}'} - \nabla^2_{\mathbf{r}}) + V_{\text{eff}}(\mathbf{r}') - V_{\text{eff}}(\mathbf{r})\right]\rho, \tag{5.1}$$

where $\rho = \rho(\mathbf{r}, \mathbf{r}')$ is the one-body density matrix and $V_{\text{eff}}(\mathbf{r})$ is an effective self-consistent potential containing interactions and external fields [36]. Using the Wigner-transform

$$f_{\text{W}}(\mathbf{r}, \mathbf{p}) = \tfrac{1}{(2\pi\hbar)^3} \int \mathrm{d}^3\mathbf{q} e^{i\mathbf{p}\dot{\mathbf{q}}} \tag{5.2}$$

Equation (5.1) can be rewritten as a transport equation

$$\tfrac{\partial}{\partial t} f_{\text{W}} + \tfrac{\mathbf{p}}{m}\nabla_{\mathbf{r}} f_{\text{W}} - \tfrac{2}{\hbar} f_{\text{W}} \sin\left(\tfrac{\hbar}{2}\overleftarrow{\nabla}_{\mathbf{p}} \cdot \overrightarrow{\nabla}_{\mathbf{r}}\right) V_{\text{eff}}(\mathbf{r}) = 0, \tag{5.3}$$

where $f_{\text{W}}(\mathbf{r}, \mathbf{p})$ has the meaning of a one-body phase-space distribution, but for the fact that it can be negative in some regions [37]. The semiclassical limit ($\hbar \to 0$) of (5.3) follows from expanding the sine in lowest order and introducing a smoothed nonnegative distribution function $f(\mathbf{r}, \mathbf{p})$. This yields the Vlasov equation

$$\tfrac{\partial}{\partial t} f + \tfrac{\mathbf{p}}{m} \cdot \nabla_{\mathbf{r}} f - \nabla_{\mathbf{p}} f \cdot \nabla_{\mathbf{r}} V_{\text{eff}}(\mathbf{r}, t) = 0. \tag{5.4}$$

Quantum effects are now solely contained in the effective potential and in the initial conditions for the distribution function. The latter we determine from the self-consistent Thomas–Fermi ground state according to

$$f^0(\mathbf{r}, \mathbf{p}) = \tfrac{2}{(2\pi\hbar)^3}\Theta(p_{\text{F}}(\mathbf{r}) - p), \tag{5.5}$$

where $p_{\text{F}}(\mathbf{r}) = \sqrt{2m[\mu - V_{\text{eff}}(\mathbf{r})]}$ is the local Fermi momentum and μ the chemical potential. Based on that, (5.4) describes the collisionless dynamics. To incorporate binary EEC, (5.4) is complemented by an Uehling–Uhlenbeck collision term [38], which results in the Vlasov–Uehling–Uhlenbeck equation

$$\tfrac{\partial}{\partial t} f + \tfrac{\mathbf{p}}{m} \cdot \nabla_{\mathbf{r}} f - \nabla_{\mathbf{p}} f \cdot \nabla_{\mathbf{r}} V_{\text{eff}}(\mathbf{r}, t) = I_{\text{UU}}. \tag{5.6}$$

The UU-collision integral reads

$$I_{\mathrm{UU}}(\mathbf{r},\mathbf{p}) = \int \mathrm{d}\Omega\, \mathrm{d}\mathbf{p}_1 \frac{|\mathbf{p}-\mathbf{p}_1|}{m} \frac{\mathrm{d}\sigma(\theta, |\mathbf{p}-\mathbf{p}_1|)}{\mathrm{d}\Omega} \times \left[f_{\mathbf{p}'} f_{\mathbf{p}_1'} (1-\tilde{f}_{\mathbf{p}})(1-\tilde{f}_{\mathbf{p}_1}) - f_{\mathbf{p}} f_{\mathbf{p}_1} (1-\tilde{f}_{\mathbf{p}'})(1-\tilde{f}_{\mathbf{p}_1'}) \right] \quad (5.7)$$

and represents a local gain–loss-balance for elastic electron–electron scattering $(\mathbf{p},\mathbf{p_1}) \leftrightarrow (\mathbf{p}',\mathbf{p_1}')$ with the energy dependent differential cross section $\mathrm{d}\sigma(\theta, |\mathbf{p}_{\mathrm{rel}}|)/\mathrm{d}\Omega$, where the scattering angle θ defines the deflection of the relative momentum vector, the local phase-space densities $f_{\mathbf{p}} = f(\mathbf{r},\mathbf{p})$, and the Pauli blocking factors in parenthesis as functions of the relative phase-space occupation for paired spins $\tilde{f}_{\mathbf{p}} = 4\pi^3\hbar^3 f_{\mathbf{p}}$. Equations (5.6) and (5.7) describe the dynamics including EEC. Note, because of the blocking factors, the collision integral vanishes in the ground state, where the Vlasov description is recovered as the limit for weak excitations. The evaluation of the UU-collision integral requires the determination of scattering cross sections, which can be obtained from quantum scattering theory. Assuming a screened electron–electron interaction potential

$$V_{\mathrm{sc}}(r) = \frac{e}{4\pi\epsilon_0} \frac{\mathrm{e}^{-r/r_{\mathrm{TF}}}}{r}, \quad (5.8)$$

where $r_{\mathrm{TF}} = (\frac{\pi}{3n_{\mathrm{e}}})^{1/6} \frac{\sqrt{a_0}}{2}$ is the Thomas–Fermi screening length for a fully degenerate Fermi gas and n_{e} is the electron density, the cross sections can be computed by partial wave analysis. For details see [39].

Within the Vlasov or VUU dynamics, a self-consistent mean-field potential of the form

$$V_{\mathrm{eff}}(\mathbf{r}) = \sum_{\mathrm{i}} V_{\mathrm{ion}}(\mathbf{r} - \mathbf{R}_i(t)) + V_{\mathrm{Har}} + V_{\mathrm{xc}} + \mathcal{E}(t) \cdot \mathbf{r} \quad (5.9)$$

is considered, containing the sum over the ion potentials for the present configuration $\mathbf{R}_i(t)$, the electron Hartree potential V_{Har}, the LDA exchange-correlation potential V_{xc} from [40], and the laser field in dipole approximation via the last term $\mathcal{E}(t) \cdot \mathbf{r}$. The Hartree term and the exchange-correlation potential are calculated from the actual total electron density $n_{\mathrm{e}}(\mathbf{r},t) = \int \mathrm{d}^3\mathbf{p}\, f(\mathbf{r},\mathbf{p},t)$. To avoid the numerically expensive propagation of strongly localized states, only valence electrons are treated explicitly in the model, while the interaction with nuclei and core electrons is described by a local pseudopotential for the sodium ions [41]. Classical motion is assumed for the ions.

In this form, the semiclassical approximation is valid for strong electronic excitations $E_{\mathrm{ex}} \gg \Delta$ and/or dense electronic energy levels $\epsilon_{\mathrm{F}} \gg \Delta$, where E_{ex} is the excitation energy, ϵ_{F} is the Fermi energy, and Δ the single particle level spacing. In particular, for sodium clusters the semiclassical method is well tested and predicts ground state geometries, optical spectra, and dynamics close to results from quantal density-functional calculations for cluster sizes $N > 10$ [35, 42, 43].

The Vlasov/VUU equation can be efficiently solved using the test particle method, which was developed in nuclear physics [37]. The key idea is to sample the continuous distribution function with a swarm of fractional particles and to map the dynamics into classical equations of motion for the discrete samples. A straightforward way of representation is

$$f(\mathbf{r}, \mathbf{p}, t) = \frac{1}{N_\mathrm{s}} \sum_i^{N_{pp}} g_r(\mathbf{r} - \mathbf{r}_i(t)) g_p(\mathbf{p} - \mathbf{p}_i(t)), \tag{5.10}$$

with the positions $\mathbf{r}_i$ and the momenta $\mathbf{p}_i$ of the test particles and the smooth weighting functions g_r and g_p in coordinate and momentum space. The parameter N_s sets the number of test particles per physical particle and defines the total number of test particles $N_{pp} = N \cdot N_s$. One possible choice for the weighting are normalized gaussians

$$g(\mathbf{x}) = \frac{1}{\pi^{3/2} d^3} \mathrm{e}^{-x^2/d^2}, \tag{5.11}$$

where d is a numerical smoothing parameter. Using the test particle ansatz, the mean-field part of the electron propagation reduces to classical motion for the test particles according to

$$\dot{\mathbf{r}}_i = \frac{\partial h_i^{pp}}{\partial \mathbf{p}_i} = \frac{\mathbf{p}_i}{m} \quad \text{and} \quad \dot{\mathbf{p}}_i = -\frac{\partial h_i^{\mathrm{pp}}}{\partial \mathbf{r}_i} = \underbrace{-\int V_{\mathrm{eff}}(\mathbf{r}) \nabla_{\mathbf{r}_i} g_r(\mathbf{r} - \mathbf{r}_i) \mathrm{d}^3\mathbf{r}}_{\mathbf{f}_i} . \tag{5.12}$$

For the semiclassical treatment, a smooth phase-space distribution is essential to suppress the tendency of classical thermalization, which requires a finite width of the test particles in practice [44]. However, this must not be a general shortcoming, as the width parameter can be used as a parameter to express a semiclassical version of the uncertainty principle (this is related to the Husimi-picture, see [45]).

As an intuitive example, the impact of resonant charging on the cluster ionization is investigated for Na_{55} (see Fig. 5.6) using the Vlasov propagation without electron–electron collisions [46]. The cluster response is simulated for excitations with 50 fs linearly polarized gaussian pulses at $\hbar\omega_\mathrm{L} = 1.54\,\mathrm{eV}$ having a peak intensity of $I_0 = 4 \times 10^{12}\,\mathrm{W\,cm^{-2}}$ for various pulse delays Δt. In the ground state, the semiclassical plasmon energy of the system is $\hbar\omega_o = 2.84\,\mathrm{eV}$, which is well above the laser photon energy. At these laser parameters, the highest cluster ionization occurs for a pulse delay of $\Delta t \approx 250\,\mathrm{fs}$.

The leading pulse causes only a weak cluster ionization (c), as the excitation is far off-resonant. This is also reflected in a small amplitude (a) of the induced dipole moment as well as its small phase shift (b) with respect to the

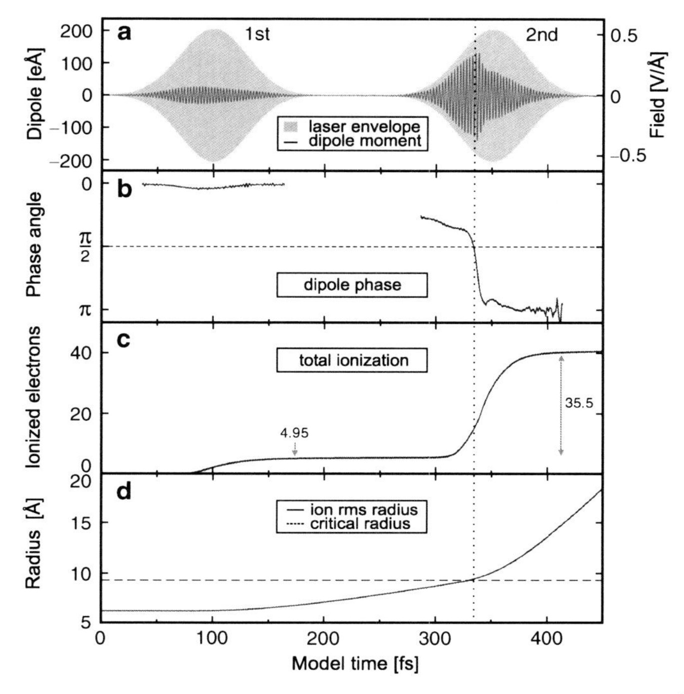

Fig. 5.6. Response of Na_{55} for dual-pulse laser excitation with $I = 4 \times 10^{12}\,\mathrm{W\,cm^{-2}}$ at 800 nm and 250 fs optical delay using the Vlasov code described in the text. Shown are the laser field envelope (*grey*) and the electron dipole amplitude (**a**), the phase lag between laser field and dipole signal (**b**), the total cluster ionization (**c**), and the root-mean-square radius of the ion distribution (**d**). Note that the dipole phase angle passes $\pi/2$ as the rms-radius is close to the critical value R_c (*dotted line*) [46]

laser field. In response to the first pulse, the excited cluster starts to expand, as can be seen from the increasing rms-cluster radius (d). A high degree of ionization is obtained when the second pulse excites the collective mode of the cluster at the critical cluster density. This value corresponds to a certain cluster size and is indicated as dashed horizontal line in Fig. 5.6d. An application of the second laser pulse near this critical value leads to an enhancement in the cluster ionization when compared to the effect of the first one (Na_{55}^{35+} vs. Na_{55}^{5+}). Strong resonant energy absorption from the second pulse is also indicated by the high dipole amplitude and the $\pi/2$-transition of the phase shift, c.f. (a) and (b).

The cluster response is modified when EEC is taken into account. The result of a comparison of Vlasov- and Vlasov-VUU simulations is given in Fig. 5.7. The response of different N (=13, 55, and 147) is calculated to show the dependence on cluster size. Resonant collective excitations at optimal pulse delays induce a pronounced enhancement in the energy deposition as well as in cluster ionization. As the total absorption and ionization are comparable

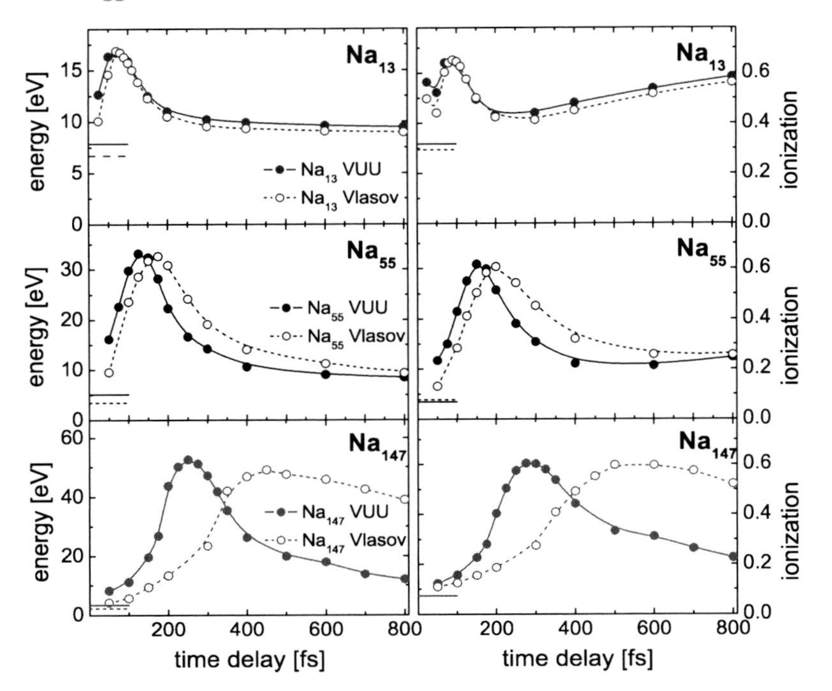

Fig. 5.7. Absorbed energy per atom (*left*) and ionization (*right*) as a function of pulse delay for different cluster sizes (Na_{13}, Na_{55}, Na_{147}). The results of the Vlasov simulations are compared to calculations where electron-electron collisions are additionally taken into account (Vlasov-VUU). Both pulses are identical (800 nm, $I = 8 \times 10^{12}$ W cm^{-2}, $\tau = 25$ fs FWHM). *Lines* at the *left* side correspond to excitation by the first pulse alone

for Vlasov and VUU, the effect of EEC on the coupling at resonance is small. But the time scales of the dynamics are strongly affected, especially for larger clusters.

From such simulations, the significance of resonance-enhanced charging can be worked out. These are model studies on simplified systems (single active electron per cluster atom), where the semiclassical method is a powerful tool to obtain detailed insight into the microscopic dynamics in the laser-cluster process.

5.4 Results and Discussion

5.4.1 Energetic Particle Emission

Ion Recoil Energy

A first and direct probe of the coupling efficiency of intense laser radiation into clusters can be achieved via a measurement of the ion recoil energy. This

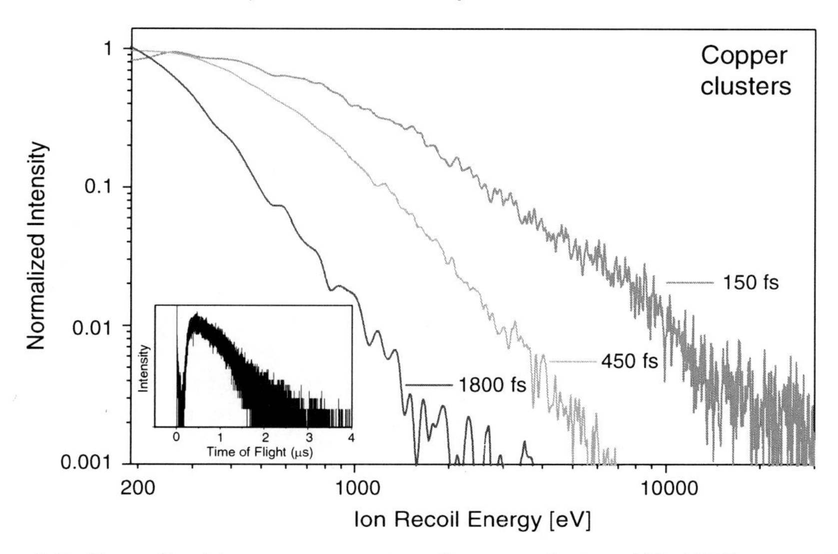

Fig. 5.8. Normalized ion energy spectra of copper clusters ($N \sim 1000$) exposed to strong laser pulses of constant fluence but different pulse durations (e.g., $1.8 \times 10^{16}\,\mathrm{W\,cm^{-2}}$ at 150 fs). The highest recoil energy is observed for the shortest pulses. Stretched pulse excitation leads to a decrease in recoil energy. Inset: one example of a time-of-flight spectrum from which the energy distribution is calculated. Adapted from [15]

is simply done by determining the flight times of energetic ions from the point of interaction to a particle detector located some tens of centimeters away. The result of such an experiment is shown in Fig. 5.8, where copper clusters produced by the magnetron gas aggregation source are exposed to 13 mJ pulses of different pulse duration. The energy spectra show broad distributions, which have maxima at low energies (several hundreds of electron-volt) but extent up to values of 30 keV. The signal stems from cluster constituents in different charge states, which are accelerated by the repulsive Coulomb potential created in the ionization. These experiments have been performed on nanometer-sized clusters and pulse intensities of about $10^{15}\,\mathrm{W\,cm^{-2}}$ at 150 fs. Increasing the cluster size and exposure with more intense pulses, the maximum energy can increase up to mega electron volt, as shown in experiments on rare gas clusters [4]. For Pb_N recoil energies up to 180 keV have been observed [47]. In general, a substantial absorption must have occurred to explain the large recoil energy values, which greatly exceed typical energies obtained in photofragmentation studies and even cluster fission experiments [48].

Several contributions influence the actual shape of the measured recoil energy distribution, which makes it difficult to extract dynamical information of the ionization process. *First,* the focusing of the laser beam leads to a large spread in the pulse intensities near the focus. This has an effect on the ionization of clusters located in different regions of the focus (*focus volume effect*). As the effective volume of a given intensity increases with the distance to the focus point, clusters located in this volume will mostly contribute to the yield, provided that the process under study can take place. Thus in an extended interaction zone, the actual contribution from the focus might be covered by a large number of events from regions of lower intensity. We have recently shown that this circumstance can be taken as an advantage. By introducing a small slit in between the interaction point and the detector the intensity dependence of the cluster charging can be studied easily by fine tuning the lens position with respect to the slit. A more detailed analysis gives good estimates for charge-resolved ionization intensity thresholds [49]. *Second*, clusters of different size are probed simultaneously due to the statistical condensation process. This might blur the information that could be extracted from the spectra. For example, assuming a uniform charge state of the cluster atoms, the Coulomb explosion will give different fragment ions recoil energy distributions depending on cluster size [50]. This is a general complication, which cannot easily be overcome without additional size selection. Both the focus volume effect as well as the broad size distribution have to be taken into account in each strong field experiment on clusters.

On the basis of these common considerations, we now return to the results shown in Fig. 5.8. In the recoil energy measurements, one cannot distinguish which ionization state contributes to the yield in different energy regions. However, it is very likely that the highest charge states will give the most energetic particles. In the measurements, different pulse durations are chosen maintaining the pulse energy. Increasing the pulse width, one measures a steady decrease in the maximum energy. We obtain this behavior also in our recoil energy studies on other metal clusters [47]. At first sight, this observation contradicts the finding of the ion charge state measurements performed under a similar laser parameter change, see, for example, the results shown in Fig. 5.2. Clearly, the maximum in the coupling efficiency at a certain but longer pulse duration – that is, relying on the outstanding contribution of the collective dipole mode in the ionization yield – is not directly visible in the recoil energy distributions. A crude estimate of the expected ion energy from a charged cluster helps to clarify this apparent discrepancy. For a homogeneously charged particle, the final energy of the cluster ions from the Coulomb explosion scales according to $\mathrm{q}^2\mathrm{n}_{\mathrm{bg}}\mathrm{r}^2 \sim q^2 r^{2/3}$, where r is the initial distance from the center, see, for example, [50]. For short pulse excitation, the clusters are nearly instantaneously ionized and only weakly charged. The final ion energy thus mainly results from the initial bulk-like atom density of the particle. At resonance, higher ion charge states are generated. However, the gain through ionization is more than compensated by the larger interatomic distances as a result of the cluster expansion. The potential energy and thus the ion recoil energy is reduced when compared to the first scenario. Therefore, the energy spectrum may not be taken as an observable to monitor delayed plasmon-enhanced ionization. Simultaneous measurements of charge state and energy of the ions are necessary to extract the details. In early attempts, a Thomson spectrometer was used by us, see Fig. 5.4 for the experimental setup. It turned out that it has to be operated by detecting only a rather small solid angle to clearly resolve the different Thomson parabola. First energy and momentum resolved distributions of exploding silver clusters have been extracted from the data [15]. We currently set up a new device based on the original B-TOF design by Lezius et al. [4]. Our improved version of the spectrometer in principle allows for a measurement of energy and momentum of *all* ion fragments in *each* laser shot.

Electron Yield

In view of resolving the ionization dynamics from the energy spectrum, the situation is much clearer when electrons are considered. The signatures in the photoemission are supposed to map more directly the efficient heating of the nanoplasma at resonance. A first estimate of the expected electron kinetic energies can be obtained from the ponderomotive interaction of an electron with a strong laser field. Because of the ponderomotive potential U_p generated by the laser field, an electron can gain kinetic energies of (3.17 U_p+IP), where IP is the atomic ionization potential. In atoms, electrons with kinetic energies up to two times U_p have been obtained [51]. The experiment on silver cluster shows a broad distribution, which extents to much higher energies, see Fig. 5.9. The yield as well as the maximum kinetic energy increases with the pulse duration, solidifying the concept of a delayed plasmon-enhanced response. A short pulse coupled into the clusters gives only values E_{kin} of 60 eV, in good agreement with results obtained on atoms, whereas a stretched pulse results in an increase in the maximum energy to values of more than 350 eV. Even more, an enhanced maximum kinetic electron energy is observed under this conditions, see Fig. 5.9, bottom. The shape of the electron energy distribution hints at electron thermalization. We, however, see in the following that such an interpretation is not sustainable when the spectra are further analyzed with respect to their angular emission characteristics.

In summarizing this section, the characteristic pulse width dependence obtained in the charge state distribution as well as in the electron energy measurements supports the presence of a resonance to be responsible for the efficient transfer of laser energy into clusters. In the next section we increase the temporal resolution to resolve finer details.

5.4.2 Time-Resolved Studies

To map the excitation dynamics in more detail, it is advantageous to apply short dual-pulses in the excitation. In doing so, one is able to probe the evolution of the system with high temporal resolution and under constant intensity conditions. Thus, one can validate the following assumption: *Can the cluster charging distinctly be separated into (1) pre-ionization to initiate*

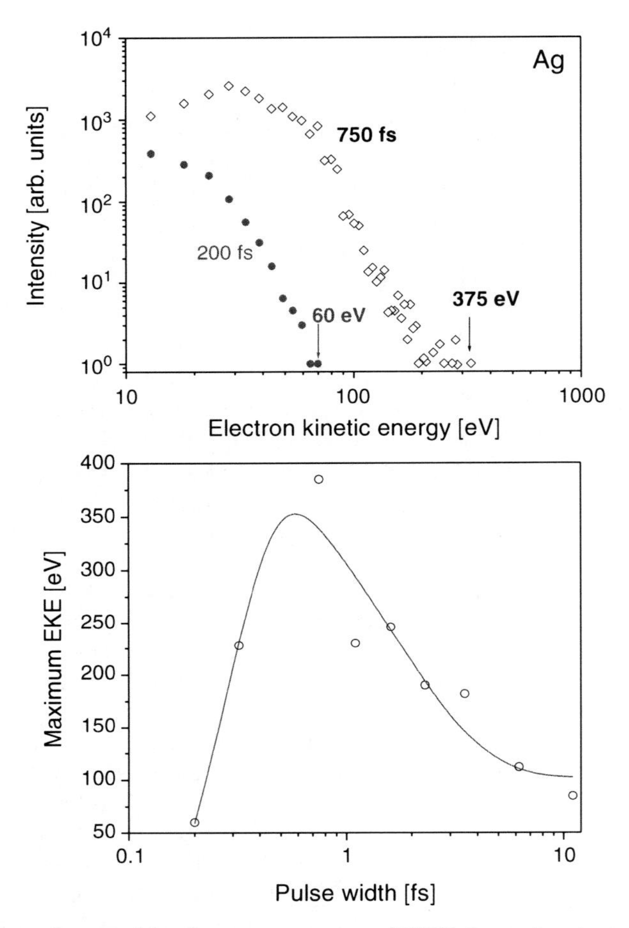

Fig. 5.9. *Top*: electron kinetic energy spectra (EKE) from Ag clusters irradiated by strong femtosecond laser pulses of different durations but constant fluence (e.g., 1.5×10^{15} W cm^{-2} at 180 fs). The maximum electron energies are marked by *arrows*. *Bottom*: the same as in the *top* figure, but depicting the maximum electron energy. The pulse widths are tuned from 200 fs up to 11 ps. The *line* serves as a guide to the eye only

the expansion followed by (2) a large energy transfer from the laser field into the particle mediated by the huge absorption cross section of the plasmon? Along this, line studies have been performed on bare and embedded silver clusters. In a droplet experiment, we used dual-pulses and recorded the yield of certain charge states as function of the optical delay. As a typical result, the intensity of Ag^{5+} is shown in Fig. 5.10, left. Surprisingly, the yield of Ag^{5+} is low when Δt is close to zero. More separating, the pulses in time leads to a substantial increase of the signal. A maximum is obtained around

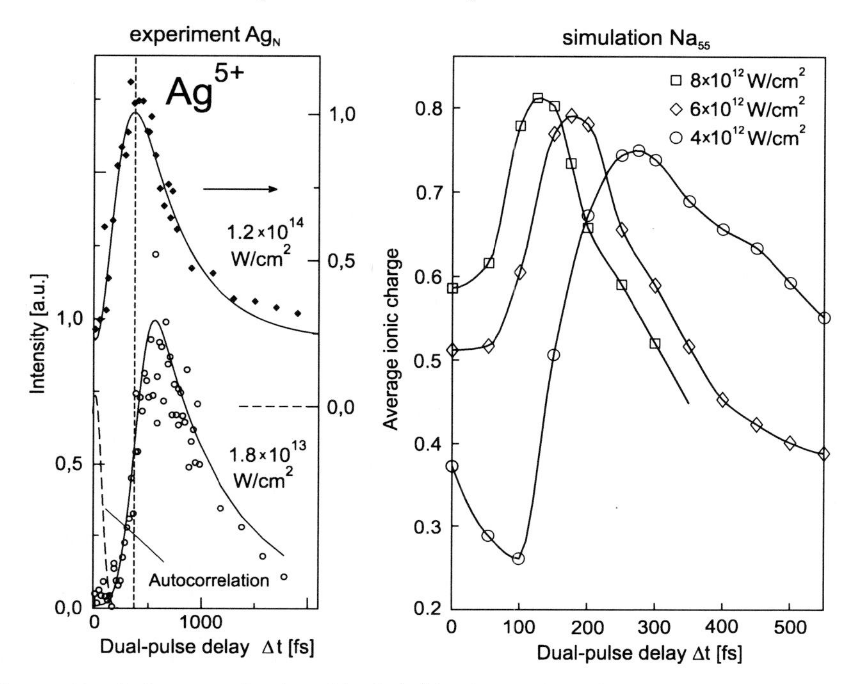

Fig. 5.10. *Left*: normalized yield of Ag^{5+} obtained in dual-pulse experiments on silver clusters (mean size 40 atoms) embedded in helium droplets under different intensity conditions. *Right*: Vlasov–MD-simulation of the average cluster charge state in Na_{55} as function of the optical delay for three different laser intensities. Qualitatively, both studies exhibit similar trends. Adapted from [46]

$\Delta t = 300$ fs and a dual-pulse effect is clearly visible for optical delays of up to 2 ps. Because of the excitation scheme, one immediately realizes that the dynamics can indeed roughly be separated into a pair of single steps, namely initial charging and delayed plasmon enhanced ionization. The impact of the degree of ionization by the leading pulse is clearly visible in the optimal time span (Δt_{M}) giving the highest ion yield. A higher pulse intensity results in a stronger initial charging and thus accelerates the expansion. This shifts the delay to smaller values, see the vertical line shown as a reference in the figure. Similar findings are obtained in Vlasov simulations, Fig. 5.10, right. The optimal delay determined in the strong field excitation of Na_{55} decrease by more than a factor of two when, for example, doubling the pulse intensity. This is accompanied by an enhanced average ionization of the cluster at resonance.

One can analyze the findings by comparing Δt_{M} for different charge states as has been done on bare nanometer-sized silver particles [52]. These are larger than the ones studied before and expand on a time scale of picosecond, which originates from the lower degree of ionization (q/atom) in the initial BSI process. Interestingly, Δt_{M} decreases with charge state q, see Fig. 5.11.

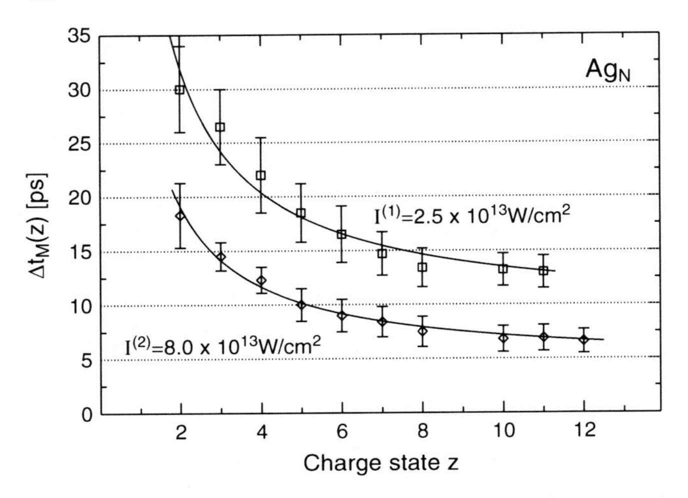

Fig. 5.11. Charge specific optimal delay Δt_{M} for dual-pulse excitation of silver particles (mean size 4.5 nm, N $\sim$ 27,000) for different pulse intensities. Exponentials are used to fit the data. Adapted from [52]

This makes sense, as clusters that are probed at optical delays larger than the time span to attain the resonance condition can only extract a lower number of photons from the laser field, resulting in a reduced charge state of the fragment ions. Again, the chosen pulse intensity has an influence on Δt_{M}, that is, a three times higher pulse intensity ($2.5 \times 10^{13}\,\mathrm{W\,cm^{-2}} \rightarrow 8.0 \times 10^{13}\,\mathrm{W\,cm^{-2}}$) reduces Δt_{M} by about a factor of two. We note that similar findings have been obtained also in experiments on clusters embedded in helium droplets [18].

The strong dependence of the yield of the multiply charged ions on the optical delay suggests similarities in the electron emission channel. Indeed, the signal of the high ionic charge states and the maximum electron energy as function of the optical delay are closely related. Figure 5.12 compares the different particle emission channels. The dual-pulse result for Ag^{10+}, which was chosen here for the comparison, resembles the findings shown before. But also the maximum electron energy behaves similar. Both find their maximum at roughly the same optical delay. This behavior is also solid under a pulse intensity variation. An increase in laser intensity shifts each curve in a similar fashion. These findings imply that (1) electrons with high kinetic energies must be present within the confining cluster potential. This has to be in the order of kilo electron-volt to allow for an efficient electron impact excitation of core electrons. Although there exits a huge outer ionization barrier, photoemission is also energetic at similar conditions, which (2) hints at a strong acceleration at resonance. In the following, the origin for the gain in energy by the electrons is analyzed in more detail.

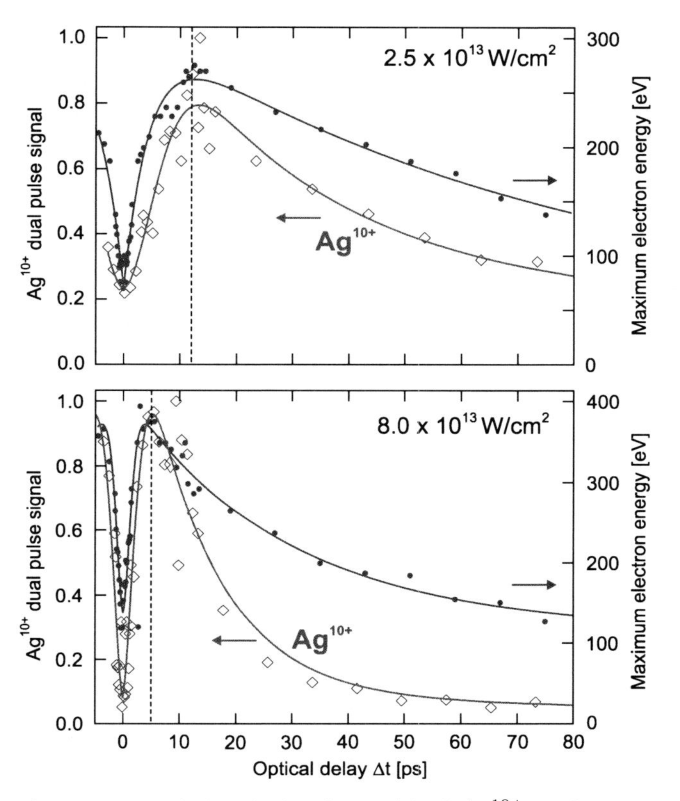

Fig. 5.12. Comparison of the dual-pulse yield of Ag^{10+} and maximum kinetic electron in the interaction of strong optical laser pulses ($2.5 \times 10^{13}\,\mathrm{W\,cm^{-2}}$ and $9.0 \times 10^{13}\,\mathrm{W\,cm^{-2}}$) with 4.5 nm ($N \sim 30{,}000$) silver particles. Both signals are clearly enhanced at the same optical delay, indicating that the electrons gain most of their kinetic energy at plasmon-enhanced absorption conditions. *Solid lines* are guides to the eyes only. Adapted from [52]

5.4.3 Directed Electron Emission

In the leading wing of the initial excitation some electrons are emitted, which causes a confining potential for the remaining electrons (inner ionization). Further ionization results in a deepening of the mean field potential, preventing the electrons more and more to overcome the outer ionization barrier, see Fig. 5.1. At first sight, it thus seems quite surprising that the highest yield in the multiply charged ions is accompanied by energetic electron emission. One could argue, that the depth of the cluster potential of several kilo electron-volt due to the high charging does not allow for the emission of energetic electrons. Evenmore, if ionization is driven by thermal emission, an isotropic pattern is expected. An angular-resolved study of the electron kinetic energy characteristics helps to solve this issue.

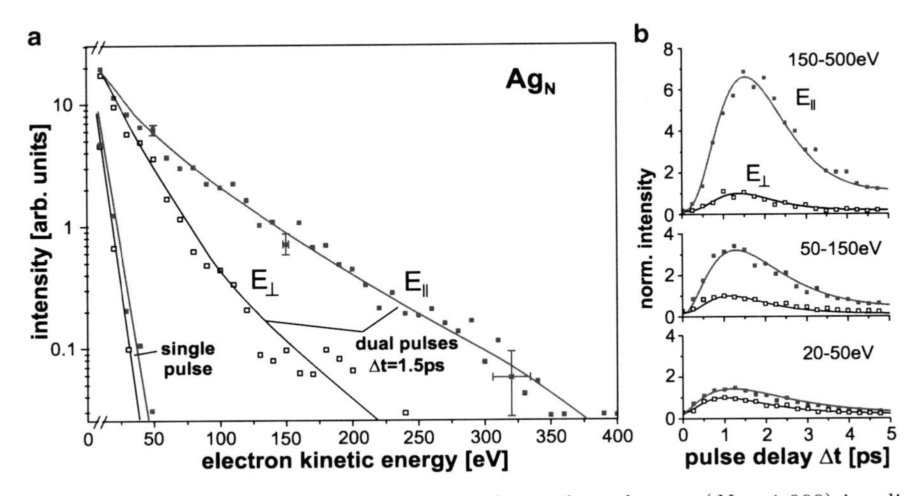

Fig. 5.13. Angular-resolved photoemission from silver clusters ($N \sim 1{,}000$) irradiated by femtosecond dual-pulses (10^{13}–10^{14} W cm^{-2}). *Left*: in the dual-pulse setup, energetic electrons are emitted preferentially in the direction of the laser polarization axis. The result of a single pulse experiment giving only low energy electrons as well as a small asymmetry in the emission is shown for comparison. *Right*: electron yield integrated over given energy windows for different optical delays Δt. A strong increase in the asymmetry is obtained when the laser field couples resonantly to the cluster. We note that the cluster size distribution is responsible for the shift in the maximum and the decrease in the asymmetry. Adapted from [53]

To experimentally resolve a possible asymmetry in the emission, electrons are detected only within a small solid angle perpendicular to the interaction plane, spanned by the direction of cluster beam and laser pulse propagation axis. Rotating a $\lambda/2$-plate, the laser polarization can be switched continuously from parallel to perpendicular orientation with respect to the direction pointing towards the electron detector. The result of a such an angular-resolved experiment is shown in Fig. 5.13, but concentrating only on the comparison between parallel and perpendicular emission with respect to the polarization axis. Almost no energetic electrons and also no clear asymmetry is present when only a single laser pulse is exposed to the clusters. Feeding in the delayed pulse at resonance the asymmetry increases, which is accompanied by a large difference in the energy of the electrons emitted in the direction of the laser polarization axis compared to the perpendicular one [53]. As the emission turns out to be strongly directed, the aforementioned possible connection of the high energy tail to a certain temperature of the strongly excited system no longer holds. Instead, the strong asymmetry points to a different process, which results in a preferential emission of energetic electrons along the laser polarization axis. Note that the laser field defines the direction of the collective electron motion, which suggests that the plasmon is responsible for the

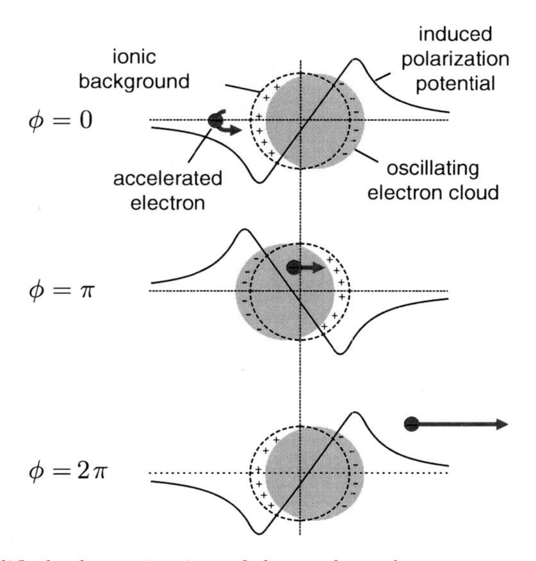

Fig. 5.14. Simplified schematic view of the *surface-plasmon assisted rescattering in clusters* (SPARC) process. The efficient acceleration at the Mie resonance is achieved via the enhanced cluster polarization field. Within a few half-cycles of the laser field, electrons gain kinetic energies in the cascade-like process. Adapted from [53]

strong and directed electron acceleration. It is worth to stress the many-body aspect of the process, which differs strongly from single electron features in the strong field excitation of atoms, like recollisions [54].

There are approaches that try to explain the acceleration of electrons in the intense laser pulse interacting with small particles [53, 55, 56]. Modeling the many-body dynamics using the Vlasov code reveals that some electrons that are out of phase with the collective mode gain most of their energy within a few cycles of the laser field. The impact of the laser-induced and resonance-enhanced polarization potential turns out to be the dynamical mechanism that accelerates the energy of the electrons. The crucial steps in the acceleration are shown schematically in Fig. 5.14. We term the mechanism *surface-plasmon assisted rescattering in clusters* (SPARC) [53]. Interestingly, the few-step acceleration leads to an attosecond pulse train of energetic electrons. Even at moderate pulse intensities of $10^{13}\,\mathrm{W\,cm^{-2}}$, peak field strengths of $35\,\mathrm{GeV\,m^{-1}}$ are attained for Na_{147}, which turns out to be five times larger than the corresponding laser field strength.

5.4.4 Control Experiments

The many different processes driving the system into a highly excited nano-plasma state calls for a procedure that optimizes the excitation.

Feedback algorithms are available, which in connection with pulse shapers (liquid crystal arrays, acousto-optical modulators, and programmable dispersive filters or deformable mirrors), see, for example, [57] for more details, can be used to modify laser pulses and to control the interaction. Prominent examples are found in molecular physics, where adaptive control is used to, for example, selectively break chemical bonds [58]. Here we apply the control scheme to the strong nonlinear response regime. The concept that is used to tailor the laser pulses was already described in Sect. 5.2. A control parameter is extracted from the recorded mass spectra, for example, the yield of the charge state under consideration is summed up and serves as a fitness value in the optimization procedure.

In one approach, silver clusters embedded in helium nanodroplets are studied. In the control experiment, the yield of selected atomic charge states are maximized. Ag^{10+} is chosen exemplarily (i.e., the yield of ions detected by the mass spectrometer) to illustrate the power of the method. Figure 5.15 shows the result of an optimization where amplitude and phase of the pulse are modified. We start with a short laser pulse of 70 fs which only gives a low yield of multiply charged ions, Fig. 5.15, top. Modifying the pulse structure using the pulse shaper, the yield of Ag^{10+} increases. In each generation, the fitness function is calculated, and after several iterations this value levels out.

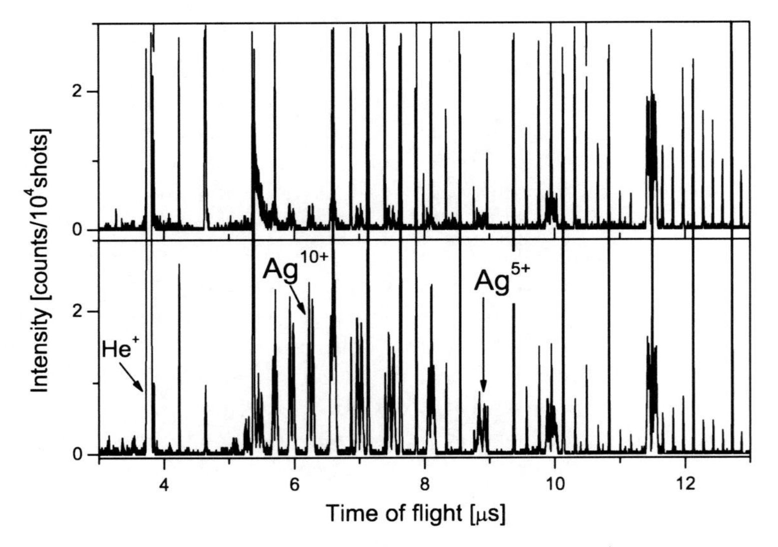

Fig. 5.15. *Top*: mass spectrum of highly charged ions (Ag^{q+}) resulting from the interaction of short pulses (70 fs FWHM) with silver clusters ($N \sim 100$) formed in helium nanodroplets. *Bottom*: optimized signal using a genetic feedback algorithm to enhance the yield of Ag^{10+}. Compared to the short pulse excitation, the signal enhances by more than one order of magnitude. The thinner lines originate from helium clusters

Typically we reach a plateau region after some tens of generations. In the mass spectrum taken under optimized pulse structure conditions, the yield of Ag^{10+} is clearly enhanced compared to the initial pulse configuration. The corresponding laser pulse leading to the higher yield analyzed by SHG-FROG shows a pulse that consists of two well-separated pulses. Finer details of the coupling process have to be worked out in the near future and may help to understand the strong field laser–matter interaction in small particles.

5.5 Conclusions

The interaction of strong optical laser pulses with metal clusters goes along with significantly enhanced absorption cross sections when compared to atoms. Because of the violent excitation, the particles completely disintegrate, resulting in atomic ions with maximum recoil energies up 180 keV. The yield of ions in high charge states is significantly enhanced when stretched or dual-pulses are used. The dependence of the highly charged ion yield on the pulse width and the dual-pulse delay gives strong evidence for the dominant role of the plasmon mode in the interaction. This is verified by Vlasov calculations on small sodium clusters, which were taken as reference model systems. The simulations give further insight into the details of the coupling dynamics and have uncovered the contribution of the plasmon in the preferred emission of energetic electrons in the direction of the laser polarization axis. Feedback controlled pulse-shaping is applied to enhance the yield of multiply charged species. In the near future, nanoplasmas might serve as pulsed sources for highly charged ions, energetic electrons, and X-rays.

References

1. T. Ditmire, J.W.G. Tisch, E. Springate, M.B. Mason, N. Hay, J.P. Marangos, M.H.R. Hutchinson, Phys. Rev. Lett. **78**, 2732–2735 (1997)
2. A. McPherson, B.D. Thompson, A.B. Borisov, K. Boyer, C.K. Rhodes, Nature **370**, 631–633 (1994)
3. T. Ditmire, J.W.G. Tisch, E. Springate, M.B. Mason, N. Hay, R.A. Smith, I. Marangos, M.M.R. Hutchinson, Nature **386**, 54–56 (1997)
4. M. Lezius, S. Dobosz, D. Normand, M. Schmidt, Phys. Rev. Lett. **80**, 261–264 (1998)
5. T. Döppner, S. Teuber, M. Schumacher, J. Tiggesbäumker, K.-H. Meiwes-Broer, Appl. Phys. B **71**, 357–360 (2000)
6. T. Ditmire, J. Zweiback, V.P. Yanovsky, T.E. Cowan, G. Hays, K.B. Wharton, Nature **398**, 489–491 (1999)
7. T. Ditmire, T. Donnelly, A.M. Rubenchik, R.W. Falcone, M.D. Perry, Phys. Rev. A **53**, 3379–3402 (1996)
8. G. Mie, Ann. Phys. **25**, 377–445 (1908)
9. L. Köller, M. Schumacher, J. Köhn, S. Teuber, J. Tiggesbäumker, K.-H. Meiwes-Broer, Phys. Rev. Lett. **82**, 3783–3786 (1999)

10. J. Tiggesbäumker, L. Köller, H.O. Lutz, K.-H. Meiwes-Broer, Chem. Phys. Lett. **190**, 42–47 (1992)
11. S. Augst, D. Strickland, D.D. Meyerhofer, S.L. Chin, J.H. Eberly, Phys. Rev. Lett. **63**, 2212–2215 (1989)
12. S. Augst, D.D. Meyerhofer, D. Strickland, S.L. Chin, J. Opt. Soc. Am. B **8**, 858–867 (1991)
13. L.V. Keldysh, Sov. Phys. JETP **20**, 1307–1314 (1965)
14. S. Zamith, T. Martchenko, Y. Ni, S.A. Aseyev, H.G. Muller, M.J.J. Vrakking, Phys. Rev. A **70**, 11201 (2004)
15. P. Radcliffe, T. Döppner, M. Schumacher, S. Teuber, J. Tiggesbäumker, K.-H. Meiwes-Broer, Contrib. Plas. Phys. **45**, 424 (2005)
16. M. Damgarten, M. Dörr, U. Eichmann, E. Lenz, W. Sandner, Phys. Rev. A **64**, 061402 (2001)
17. E.M. Snyder, S.A. Buzza, A.W. Castleman Jr., Phys. Rev Lett. **77**, 3347–3350 (1996)
18. T. Döppner, Th. Diederich, A. Przystawik, N.X. Truong, Th. Fennel, J. Tiggesbäumker, K.-H. Meiwes-Broer, Phys. Chem. Chem. Phys. **9**, 4639–4652 (2007)
19. W.A. de Heer, Rev. Mod. Phys. **65**, 611–676 (1993)
20. W. Schulze, B. Winter, I. Goldenfeld, J. Chem. Phys. **87**, 2402–2403 (1987)
21. H. Haberland, M. Mall, M. Moseler, Y. Qiang, T. Reiners, Y. Thurner, J. Vac. Sci. Technol. A **12**(5), 2925 (1994)
22. T.G. Dietz, M.A. Duncan, D.E. Powers, R.E. Smalley, J. Chem. Phys. **74**, 6511 (1981)
23. H.R. Siekmann, Ch. Lüder, J. Faehrmann, H.O. Lutz, K.-H. Meiwes-Broer, Z. Phys. D **20**, 417–420 (1991)
24. J. Passig, K.-H. Meiwes-Broer, J. Tiggesbäumker, Rev. Sci. Instr. **77**, 093304 (2006)
25. Th. Diederich, T. Döppner, Th. Fennel, J. Tiggesbäumker, K.-H. Meiwes-Broer, Phys. Rev. A **72**, 023203 (2005)
26. J. Tiggesbäumker, F. Stienkemeier, Phys. Chem. Chem. Phys. **9**, 4748–4770 (2007)
27. B.A. Mamyrin, Int. J. Mass Spec. Ion Proc. **131**, 1–19 (1994)
28. P. Kruit, F.H. Read, J. Phys. E. **16**, 313 (1983)
29. P. Maine, D. Strickland, P. Bado, M. Pessot, G. Mourou, J. Quant. Electr. **24**, 398 (1988)
30. P. Brumer, M. Shapiro, Sci. Am. **3**, 34 (1995)
31. D.J. Tannor, R. Kosloff, S.A. Rice, J. Chem. Phys. **84**, 5805–5820 (1986)
32. P. Tournois, Opt. Comm. **140**, 245–249 (1997)
33. R. Trebino, K.W. DeLong, D.N. Fittinghoff, J.N. Sweetser, M.A. Krumbügel, B.A. Richman, D.J. Kane, Rev. Sci. Inst. **68**(9), 3277 (1997)
34. U. Saalmann, C. Siedschlag, J.M. Rost, J. Phys. B **39**, R39 (2006)
35. P.G. Reinhard, E. Suraud, *Introduction to Cluster Dynamics* (Wiley, New York, 2003)
36. Th. Fennel, J. Köhn, in *Semiclassical Description of Quantum Many-Particle Dynamics in Strong Laser Fields*, ed. by H. Fehske, R. Schneider, A. Weisse. Lecture Notes in Physics: Computational Many-Particle Physics (Springer, Berlin, 2008), pp. 255–273
37. G.F. Bertsch, S. Das Gupta, Phys. Rep. **160**, 191 (1988)

38. E.A. Uehling, G.E. Uhlenbeck, Phys. Rep. **43**, 552 (1933)
39. J. Köhn, R. Redmer, K.-H. Meiwes-Broer, T. Fennel, Phys. Rev. A **77**, 033202 (2008)
40. O. Gunnarsson, B.I. Lundquist, Phys. Rev. B **13**, 4274 (1976)
41. Th. Fennel, K.-H. Meiwes-Broer, G.F. Bertsch, Eur. Phys. J. D **29**, 367–378 (2004)
42. L. Plagne, J. Daligault, K. Yabana, T. Tazawa, Y. Abe, C. Guet, Phys. Rev. A **61**, 33201 (2000)
43. E. Suraud C. Legrand, P.-G. Reinhard, J. Phys. B **39**, 2481 (2006)
44. C. Jarzynski, G.F. Bertsch, Phys. Rev. C **53**(2), 1028–1031 (1996)
45. A. Domps, P.-G. Reinhard, E. Suraud, Ann. Phys **260**, 171 (1997)
46. T. Döppner, Th. Fennel, Th. Diederich, J. Tiggesbäumker, K.-H. Meiwes-Broer, Phys. Rev. Lett. **94**, 013401 (2005)
47. S. Teuber, T. Döppner, T. Fennel, J. Tiggesbäumker, K.-H. Meiwes-Broer, Eur. Phys. J. D **16**, 59–64 (2001)
48. U. Näher, S. Bjornholm, S. Frauendorf, F. Garcias, C. Guet, Phys. Rep. **285**(6), 245–320 (1997)
49. T. Döppner, J.P. Müller, A. Przystawik, J. Tiggesbäumker, K.-H. Meiwes-Broer, Eur. Phys. J. D **43**, 261 (2007)
50. I. Last, J. Jortner, J. Chem. Phys. **121**, 3030–3043 (2004)
51. B. Walker, B. Sheehy, L.F. DiMauro, P. Agostini, K.J. Schafer, K.C. Kulander, Phys. Rev. Lett. **73**, 1227–1230 (1994)
52. T. Döppner, Th. Fennel, P. Radcliffe, J. Tiggesbäumker, K.-H. Meiwes-Broer, Eur. Phys. J. D **36**, 165–171 (2005)
53. Th. Fennel, T. Döppner, J. Passig, Ch. Schaal, J. Tiggesbäumker, K.-H. Meiwes-Broer, Phys. Rev. Lett. **98**, 143401 (2007)
54. P.B. Corkum, Phys. Rev. Lett. **71**(13), 1994–1997 (1993)
55. V. Kumarappan, M. Krishnamurthy, D. Mathur, Phys. Rev. A **67**(4), 043204 (2003)
56. U. Saalmann, J.M. Rost, Phys. Rev. Lett. **100**, 133006 (2008)
57. A.M. Weiner, Rev. Sci. Instr. **71**, 1929–1960 (2000)
58. A. Assion, T. Baumert, M. Bergt, T. Brixner, B. Kiefer, V. Seyfried, M. Strehle, G. Gerber, Science **282**, 919–921 (1998)

6

Mott Effect in Nuclear Matter

Gerd Röpke

Abstract. Nuclear matter consisting of protons and neutrons is an interesting strongly interacting quantum system featuring many-body effects. The equation of state (EoS) of nuclear matter at finite temperature and density with various proton fractions is considered, in particular the region of medium excitation energy given by the temperature range $T \leq 30\,\text{MeV}$ and the baryon density range $\rho_\text{B} \leq 10^{14.2}\,\text{g}\,\text{cm}^{-3}$. In this region ("warm and dilute asymmetric nuclear matter"), the formation of few-body correlations, in particular bound clusters, has to be taken into account. Based on a many-particle Green function approach, the medium modification of the clusters is described by self-energy and Pauli blocking effects, and the cluster-mean field approximation is given. These medium effects lead to a shift of the binding energies as well as a modification of scattering properties. Because of the shift, bound states will merge with the continuum of scattering states at increasing density and are dissolved (Mott effect). Results of the Mott effect for different nuclei embedded in nuclear matter are given.

Thermodynamic properties are influenced by the formation and dissolution of bound states. The nuclear matter EoS is considered within a generalized Beth–Uhlenbeck approach. The connection with the Brueckner Hartree–Fock and Relativistic Mean-Field theory is outlined. Benchmarks such as virial expansion in the low-density limit or the low-temperature limit are considered. An interesting effect is the formation of a two-nucleon quantum condensate, showing the crossover from Cooper pairing to Bose–Einstein condensation. Correlations in the condensate such as quartetting are an interesting issue. The structure of the quantum condensate is determined by the existence of bound states and the Mott effect.

The resulting thermodynamic properties, incorporating the Mott effect, are of interest for heavy-ion collisions and astrophysical applications. The Mott effect is also of relevance for the structure of finite nuclei, especially dilute excited states like the Hoyle state of ^{12}C.

6.1 Introduction

The equation of state (EoS), the composition, and the possible occurrence of phase transitions in nuclear matter are not only the widely discussed topics in nuclear theory [1], but also of great interest in astrophysics and cosmology.

Experiments on heavy ion collisions, performed over the last decades, have given new insight into the behavior of nuclear systems in a broad range of densities and temperatures. The observed cluster abundances, their spectral distribution, and correlations in momentum space can deliver information about the state of dense, highly excited matter. In a recent review [2], constraints to the EoS have been investigated. Different versions have been considered in the high-density region, and comparison with the properties of neutron star may discriminate between more or less reasonable equations of state.

We restrict ourselves to matter in equilibrium at temperatures $T \leq 30\,\mathrm{MeV}$ and baryon number densities $n_\mathrm{B} \leq 0.2\,\mathrm{fm}^{-3}$, where the quark substructure and the excitation of internal degrees of freedom of the nucleons (protons p and neutrons n) are not of relevance and the nucleon–nucleon interaction can be represented by an effective interaction potential. In this region of the temperature-density plane, we investigate how the quasi-particle picture will be improved if few-body correlations are taken into account. The influence of cluster formation on the EoS is calculated for different situations, and the occurrence of phase instabilities is investigated. Derived from the full spectral function, the concept of composition will be introduced as an approximation to describe correlations in dense systems. Another interesting issue is the formation of quantum condensates.

A quantum statistical approach to the thermodynamic properties of nuclear matter can be given using the method of thermodynamic Green functions [3]. In general, within the grand canonical ensemble, the EoS $n_\tau(T, \mu_{\tau'})$ relates the particle number densities n_τ to the temperature (T) and the chemical potentials μ_τ of protons (p) or neutrons (n), where the internal quantum number τ can be introduced to describe besides isospin (p, n) also spin and further quantum numbers. This EoS is obtained from the single-particle spectral function, which can be expressed in terms of the self-energy. Then, thermodynamic potentials such as the pressure $p(T, \mu_\tau)$ or the density of free energy $f(T, n_\tau)$ are obtained by integrations. From these thermodynamic potentials, all other equilibrium thermodynamic properties can be derived. In particular, the stability of the homogeneous system against phase separation has to be considered.

The main quantity to be evaluated is the self-energy. Different approximations can be obtained by partial summations within a diagrammatic representation. The formation of bound states is taken into account considering ladder approximations [4], leading in the low-density limit to the solution of the Schrödinger equation. The effects of the medium can be included in a self-consistent way within the cluster-mean field approximation (see [5] for references), where the influence of the correlated medium on the single particle states as well as on the clusters is considered in first order with respect to the interaction. As a point of significance, the single particle and the bound states are considered on equal footing. Besides single-particle self-energy shifts of the constituents, the bound state energies are also modified by the Pauli blocking due to the correlated medium. An extended discussion of the two-particle problem can be found in [6]. With a generalized Beth–Uhlenbeck formula, not only the two-particle properties such as deuteron formation and scattering phase shifts have been used to construct a nuclear matter EoS, but also the influence of the medium has been taken into account, which leads to the suppression of correlations at high densities. Similarly, also three- and four-particle bound states can be included [4], and also the medium-dependent shift of the cluster binding energies was investigated, see [7].

If a singularity in the bosonic (even A) medium-modified few-body T-matrix T_A arises at energy that coincides with the corresponding chemical potential $Z\mu_p + (A - Z)\mu_n$, then the formation of a quantum condensate is indicated. Different kinds of quantum condensates have been considered in nuclear matter [8, 9]. They become obvious if the binding energy of nuclei is investigated [10]. Correlated condensates are found to give a reasonable description of near-threshold states of $n\alpha$ nuclei [11]. The contribution of condensation energy to the nuclear matter EoS would be of importance and has to be taken into account not only in mean-field approximation but also considering correlated condensates.

The EoS can be applied to different situations. In astrophysics, the relativistic EoS of nuclear matter was investigated recently [12] for supernova explosions. If nuclei are considered as inhomogeneous nuclear matter, within a local density approximation the EoS can serve for comparison to estimate the role of correlations. In nuclear reactions a nonequilibrium theory is needed, but within a simple approach such as the freeze out concept or the coalescence model, the results from the EoS may be used to describe heavy ion reactions.

6.2 Single Particle Spectral Function and Self-Energy

A quantum statistical approach to nuclear matter at low densities and temperatures can be given in the nonrelativistic case starting from a Hamiltonian

$$H = \sum_1 E(1) a_1^\dagger a_1 + \sum_{121'2'} V(12,1'2') a_1^\dagger a_2^\dagger a_{2'} a_{1'}, \tag{6.1}$$

where we can use linear momentum, spin, and isospin to characterize the single-nucleon state, $\{1\} = \{\boldsymbol{p}_1, \sigma_1, \tau_1\}$, so that the kinetic energy is given by $E(1) = p_1^2/2m_1$, while the interaction potential $V(12, 1'2')$ can be given in the different spin and isospin channels. We also use the charge number $Z_1 = 1$ for protons ($\tau_1 = p$) and $Z_1 = 0$ for neutrons ($\tau_1 = n$).

The basic problem of a many-particle approach to nuclear matter is the lack of a first principle derivation of the interaction potential from QCD. Instead, the potential can be introduced empirically reproducing measured quantities of the two-particle system such as scattering phase shifts and binding energies. This way, standard parametrization for the nucleon–nucleon interaction like the Bonn or Paris potentials has been introduced, and for the sake of solution of the Schrödinger equation, the separable forms like PEST4 are more appropriate [13]. Mention that local potentials can be represented by a sum of separable potentials, so that both can be considered as equivalent. Furthermore, they contain the full information about scattering phase shifts in the parameter region of relevance.

Another remark refers to the nonrelativistic description. As long as relativistic effects can be neglected, the use of concepts like the Hamiltonian, bound state wave functions, and scattering phase shifts allow for a systematic treatment of correlations in the many-particle system. However, at densities above saturation and temperatures comparable with the nucleon rest mass, a relativistic description is indispensable.

A quantum statistical approach can be given using the Green function formalism. The treatment of nuclear matter with finite temperature Green functions, including the formation of bound states, has been given in [4, 6], and the EoS for nuclear matter has been deduced. It is most appropriate to start with the EoS for nucleon densities $n_\tau(\beta, \mu_p, \mu_n)$ as a function of temperature $T = 1/(k_\mathrm{B}\beta)$ and chemical potentials μ_τ for protons and neutrons, respectively. With the spectral function $A(1, \omega; \beta, \mu_p, \mu_n)$, we have

$$n_{\tau_1}(\beta, \mu_p, \mu_n) = \frac{2}{\Omega} \sum_p \frac{1}{2\pi} \int_{-\infty}^{\infty} f_{1,Z_1}(\omega; \beta, \mu_p, \mu_n) A(1, \omega; \beta, \mu_p, \mu_n)\, \mathrm{d}\omega, \quad (6.2)$$

where Ω is the system volume, and summation over spin direction is collected in the factor 2. The Fermi function is $f_{1,Z_1}(\omega) = [\exp(\beta(\omega - \mu_1)) + 1]^{-1}$ and depends on the inverse temperature β and the chemical potential $\mu_1 = \mu_p$ for $\tau_1 = p$, that is, $Z_1 = 1$, or $\mu_1 = \mu_n$ for $\tau_1 = n$, that is, $Z_1 = 0$.

At this point, we consider the densities of both protons and neutrons or, alternatively, the baryon density $n_\mathrm{B} = (n_n + n_p)$ and the asymmetry of nuclear matter $\alpha = (n_n - n_p)/(n_n + n_p)$ as given parameters. In a further evaluation, allowing for weak interaction, β equilibrium may be achieved which is of interest, for example, for astrophysical applications. Besides the frozen equilibrium where n_p, n_n are fixed, we assume homogeneity in space as a first step. For given T, Ω, and particle numbers $N_\tau = n_\tau \Omega$, the minimum of the free energy $F = \mathcal{F}\Omega$ has to be found. This thermodynamic potential follows from integration, for example, $\mathcal{F}(T, n_p, n_n) = \int_0^{n_n} \mu_n(T, 0, n_n')\, \mathrm{d}n_n' + \int_0^{n_p} \mu_p(T, n_p', n_n)\, \mathrm{d}n_p'$. For stability against phase separation, the curvature matrix $\mathcal{F}_{\tau,\tau'} = \partial\mathcal{F}/\partial n_\tau \partial n_{\tau'}|_T$ has to be positive, that is, $\mathrm{Tr}\,[\mathcal{F}_{\tau,\tau'}] \geq 0$, $\mathrm{Det}\,[\mathcal{F}_{\tau,\tau'}] \geq 0$.

The basic equations of a Green function approach to the EoS can be found in different papers, see [6]. We give here only some final results. The spectral function is related to the self-energy according to

$$A(1, \omega) = \frac{2\mathrm{Im}\Sigma(1, \omega - i0)}{(\omega - E(1) - \mathrm{Re}\Sigma(1, \omega))^2 + (\mathrm{Im}\Sigma(1, \omega - i0))^2}, \quad (6.3)$$

where the imaginary part has to be taken for a small negative imaginary part in the frequency.

For the self-energy Σ, a cluster decomposition [4] is possible. The diagrams are calculated as

$$\Sigma(1, z_\nu) = \sum_A \sum_{\Omega_\lambda, 2...A} G^{(0)}_{(A-1)}(2, ..., A, \Omega_\lambda - z_\nu) T_A(1...A, 1'...A', \Omega_\lambda) \quad (6.4)$$

with the free $(A-1)$ (quasi-) particle propagator

$$G^{(0)}_{(A-1)}(2, ..., A, z) = \frac{1}{z - E_2 - ... - E_A} \frac{f_{1,Z_2}(2)...f_{1,Z_A}(A)}{f_{A-1,Z_{A-1}}(E_2 + ... + E_A)} \quad (6.5)$$

and

$$f_{A,Z}(E) = [\exp((E - Z\mu_p - (A - Z)\mu_n)/k_B T) - (-1)^A]^{-1}. \quad (6.6)$$

The T_A matrices are related to the A-particle Green functions

$$\begin{aligned} &\mathrm{T}_A(1 \ldots A, 1' \ldots A', z) = V_A(1 \ldots A, 1' \ldots A') \\ &+ V_A(1 \ldots A, 1'' \ldots A'') G_A(1'' \ldots A'', 1''' \ldots A''', z) V_A(1''' \ldots A''', 1' \ldots A'), \end{aligned} \quad (6.7)$$

with the potential $V_A(1 \ldots A, 1' \ldots A') = \sum_{i<j} V(ij, i'j') \prod_{k \neq i,j} \delta_{k,k'}$, and subtraction of double counting diagrams when inserting the T matrices into the self-energy. The solution of the A-particle propagator in the low-density limit is given by (we include the isospin quantum number Z in the internal quantum number $\nu = Z, n$)

$$G_A(1 \ldots A, 1' \ldots A', z) = \sum_{\nu, P} \frac{\psi_{A,\nu,P}(1 \ldots A) \psi^*_{A,\nu,P}(1' \ldots A')}{z - E_{A,\nu,P}} \quad (6.8)$$

using the eigenvalues $E_{A,\nu,P}$ and the wave functions $\psi_{A,\nu,P}(1 \ldots A)$ of the A-particle Schrödinger equation, P denotes the total momentum, and the internal quantum number ν covers bound as well as scattering states.

The evaluation of the EoS in the low-density limit is straightforward. With

$$\mathrm{T}_A(1 \ldots A, 1' \ldots A', z) = \sum_{\nu, P} \quad (6.9)$$
$$\frac{(z - E_1 - ... - E_A) \psi_{A,\nu,P}(1 \ldots A) \psi^*_{A,\nu,P}(1' \ldots A')(E_{A,\nu,P} - E_{1'} - ... - E_{A'})}{z - E_{A,\nu,P}}$$

we can perform the Ω_λ summation in (6.4). We obtain the result

$$\begin{aligned} &\sum_{\Omega_\lambda} \frac{1}{\Omega_\lambda - z_\nu - E_2 ... - E_A} \frac{(\Omega_\lambda - E_1 ... - E_A)(E_{A,\nu,P} - E_1 ... - E_A)}{\Omega_\lambda - E_{A,\nu,P}} \\ &= f_{A-1}(E_2 + ... + E_A) \frac{z_\nu - E_1}{z_\nu + E_2 + ... + E_A - E_{A,\nu,P}} \\ &\quad - f_A(E_{A,\nu,P}) \frac{E_1 + ... + E_A - E_{A,\nu,P}}{z_\nu + E_2 + ... + E_A - E_{A,\nu,P}}. \end{aligned} \quad (6.10)$$

Taking $\mathrm{Im}\,\Sigma$ and integrating the δ function arising from the pole in the denominator, we have the leading term in density given by

$$f_1(E_{A,\nu,P} - E_2 - ... - E_A) f_{A-1}(E_2 + ... + E_A) = f_A(E_{A,\nu,P}).$$

Considering only the bound-state contributions, we have the result for the EoS in the low-density, low-temperature limit

$$n_{\mathrm{B}}(\beta, \mu_p, \mu_n) = \Omega^{-1} \sum_{A,\nu,P} A\, f_{A,Z}(E_{A,\nu,P}), \tag{6.11}$$

which is an *ideal mixture of components* obeying Fermi or Bose statistics.

In the classical limit, the integrals over P can be carried out, and one obtains the mass-action law that determines the matter composition at given temperature and total particle density. At low temperatures, quantum effects become relevant. The most dramatic is Bose–Einstein condensation (BEC), which occurs for the channels with even A when $E_{A,\nu,P} - Z\mu_p - (A-Z)\mu_n = 0$. As the temperature is decreased from relatively high values toward zero, BEC occurs first for those clusters with largest binding energy per nucleon. If we take into account also the formation of α clusters in nuclear matter, the two-particle (deuteron) binding energy per nucleon is 1.11 MeV, while the four-particle (α) binding energy per nucleon is 7 MeV. One therefore expects that a quantum condensate of α particles is formed.

The Green function approach provides us with a systematic treatment of interaction in nuclear matter. Thus, we can include the contribution of scattering to the EoS. Furthermore, the properties of the bound states are changed due to the influence of the medium as discussed later.

6.3 Two-Particle Contribution: Generalized Beth–Uhlenbeck Formula and Virial Expansion

For the self-energy, appropriate approximations have to be performed. As a trivial case, the free fermion gas approximation follows for vanishing self-energy. Here we discuss first the inclusion of two-particle correlations as given by the two-nucleon T matrix (or the so-called Brueckner G matrix) [6]. We show how a self-consistent approach can be given, which includes medium effects, so that not only the second virial coefficient is obtained in the low-density limit, but also the fading of correlations at higher densities is described.

In the so-called ladder approximation, which is known as the Bethe–Goldstone equation, we have (see (6.7))

$$\mathrm{T}_2^L(12,34,z) = V(1234) + \sum_{5678} V(12,56) G_2^0(56,78,z) \mathrm{T}_2^L(78,34,z). \tag{6.12}$$

The propagator of the noninteracting pair within a dense medium reads

$$G_2^0(12,34,z) = \int \frac{\mathrm{d}\omega}{2\pi} \int \frac{\mathrm{d}\omega'}{2\pi} \frac{1 - f_{1,Z_1}(\omega) - f_{1,Z_2}(\omega')}{\omega + \omega' - z} A(1,\omega)A(2,\omega')\delta_{13}\delta_{24}, \tag{6.13}$$

where we took the full single-particle propagator as given by the spectral function. As given earlier (6.5) in the quasiparticle approximation, the spectral function becomes δ-like and the integrals over ω, ω' can be performed. The factor $(1 - f_1 - f_1)$ (Feynman–Galitsky) results from the Matsubara sum. It accounts for the Pauli principle, which forbids the scattering into final states that are occupied by surrounding particles. In contrast to the Brueckner G matrix, which has the Pauli blocking as $(1 - f_1)(1 - f_1)$, the hole contributions are included. This is necessary, for example, to obtain the transition to superfluidity, see Sect. 6.11.

Constructing the self-energy in the approximation $\Sigma_2^L = \mathrm{T}_2^L G$, we have (see [6])

$$\begin{aligned} \Sigma_2^L(1,z) = \sum_2 \int \frac{\mathrm{d}E}{2\pi} A(2,E) \Big[& f_{1,Z_2}(E)(V(12,12) - \mathrm{ex}) \\ & - \int \frac{\mathrm{d}E'}{2\pi} (\mathrm{Im}\mathrm{T}_2^L(12,12,E'+i0) - \mathrm{ex}) \frac{f_{1,Z_2}(E) + f_{2,Z}(E')}{E' - E - z} \Big] . \end{aligned} \tag{6.14}$$

To obtain the so-called chemical picture, an expansion with respect to small imaginary part of the self-energy will be performed. Introducing the quasiparticle energies according to

$$E^{\mathrm{qu}}(1) = e(1) = E(1) + v(1) \approx v(1, p_1 = 0) + \frac{p_1^2}{2m_1^*} \tag{6.15}$$

with $v(1) = \mathrm{Re}\,\Sigma(1, e(1))$, we have

$$\begin{aligned} A(1,E) = & \frac{2\pi\delta(E - e(1))}{1 - ((\mathrm{d}/\mathrm{d}z)\mathrm{Re}\,\Sigma(1,z)|_{z=e(1)}} \\ & -2\mathrm{Im}\,\Sigma(1, E + i0) \frac{\mathrm{d}}{\mathrm{d}E} \frac{\mathcal{P}}{E - e(1)} . \end{aligned} \tag{6.16}$$

The spectral function consists of the δ-shaped contribution of free quasiparticles and a correlation contribution. Expanding the denominator of the first term and using the spectral representation of the self-energy [6], we arrive at the decomposition $n_{\mathrm{B}}(T, \mu_p, \mu_n) = n_{\mathrm{free}}(T, \mu_p, \mu_n) + 2n_{\mathrm{corr}}(T, \mu_p, \mu_n)$ of the nucleon density as a function of chemical potential and temperature, with

$$n_{\mathrm{free}}(T, \mu_p, \mu_n) = \Omega^{-1} \sum_1 f_{1,Z_1}(E^{\mathrm{qu}}(1)) \tag{6.17}$$

and

$$
\begin{aligned}
n_{\mathrm{corr}}(T,\mu_p,\mu_n) &= \Omega^{-1}\sum_{12}\int\frac{\mathrm{d}E}{2\pi}f_{2,Z}(E)\sum_{12}[1-f_{1,Z_1}(e(1))-f_{1,Z_2}(e(2))] \\
&\times\left[(\mathrm{Im\,T}(12,12,E+i0)-\mathrm{ex})\frac{\mathrm{d}}{\mathrm{d}E}\frac{\mathcal{P}}{e(1)+e(2)-E}\right. \\
&\left.-\pi\delta(E-e(1)-e(2))\frac{\mathrm{d}}{\mathrm{d}E}(\mathrm{Re\,T}(12,12,E)-\mathrm{ex})\right]. \quad (6.18)
\end{aligned}
$$

Compared with the Brueckner–Bethe–Goldstone approach, the density contribution of correlated pairs of quasiparticles $n_{\mathrm{corr}}(T,\mu_p,\mu_n)$ is new. As shown in [6], it can be expressed in terms of on-shell T-matrix elements, which are represented via generalized scattering phases affected by the nuclear medium defined in the different two-particle channels τ specifying the spin-triplet (isospin-singlet) channel $\tau={}^3S_1$ and the spin-singlet (isospin-triplet) channel $\tau={}^1S_0$. We give the final expression for the generalized Beth–Uhlenbeck equation

$$
\begin{aligned}
n_{\mathrm{corr}}(T,\mu_p,\mu_n) &= 3\Omega^{-1}\sum_{P>P_{\mathrm{d}}^{\mathrm{Mott}}}[f_{2,1}(E_{\mathrm{cont}}(P)+E_{\mathrm{d}}(P))-f_{2,1}(E_{\mathrm{cont}}(P))] \\
-\sum_{P,\tau}g_\tau\int_0^\infty\frac{\mathrm{d}E}{\pi}&\left[\frac{\mathrm{d}}{\mathrm{d}E}f_{2,Z_\tau}(E_{\mathrm{cont}}(P)+E)\right](\delta_\tau(E)-\sin\delta_\tau(E)\cos\delta_\tau(E)).
\end{aligned}
\quad (6.19)
$$

Here, $E_{\mathrm{cont}}(P;T,\mu_p,\mu_n)=P^2/4m+2v(P/2;T,\mu_p,\mu_n)$ denotes the continuum edge for a pair with total momentum P. The medium dependent scattering phase shifts $\delta_\tau(E;T,\mu_p,\mu_n)$ in the channel τ, degeneration factor g_τ, and the medium dependent deuteron binding energy $E_{\mathrm{d}}(P;T,\mu_p,\mu_n)$ follow from the solution of the T-matrix (6.12), which also determines the medium-modified bound state energies. The chemical potential $\mu_\tau=\mu_1+\mu_2$ denotes the sum of the chemical potentials of the nucleons in the channel τ. The in-medium scattering phase shifts for the channel τ at the energy of relative motion E follow from the T matrix for a given interaction as will be shown later on. The minimum center of mass momentum of a proton–neutron pair above which a deuteron-like bound state can be formed is denoted by $P_{\mathrm{d}}^{\mathrm{Mott}}(T,\mu_p,\mu_n)$. Details are given below in Sect. 6.7.

Before evaluating these expressions for a given nucleon–nucleon potential, we present the ordinary Beth–Uhlenbeck equation. In the low-density limit, all medium corrections to the single and two-particle properties can be neglected. Furthermore, at fixed temperature, Fermi and Bose functions can be expanded near the classical distribution if $n_{\mathrm{B}}\Lambda^3/4\ll 1$, where $\Lambda=(2\pi\hbar^2/mk_{\mathrm{B}}T)^{1/2}$ is the thermal wavelength of nucleons. The ordinary Beth–Uhlenbeck formula is reproduced, with up to second order in the density

$$n_n = \frac{2}{\Lambda^3} e^{\mu_n/k_B T} + \frac{4}{\Lambda^3} e^{2\mu_n/k_B T} b_n + \frac{4}{\Lambda^3} e^{(\mu_n+\mu_p)/k_B T} b_{pn}, \tag{6.20}$$

$$n_p = \frac{2}{\Lambda^3} e^{\mu_p/k_B T} + \frac{4}{\Lambda^3} e^{2\mu_p/k_B T} b_p + \frac{4}{\Lambda^3} e^{(\mu_n+\mu_p)/k_B T} b_{pn}. \tag{6.21}$$

Taking the relative energy E, we have

$$b_n = b_p = \frac{2}{2^{1/2}\pi k_B T} \int_0^\infty e^{-E/k_B T} \delta_{^1S_0}(E)\, dE - 2^{-5/2}, \tag{6.22}$$

$$b_{pn} = \frac{3}{2^{1/2}} \left[e^{-E_d/k_B T} - 1 + \frac{1}{\pi k_B T} \int_0^\infty e^{-E/k_B T} \delta_{^3S_1}(E)\, dE \right] + \frac{1}{2^{1/2}\pi k_B T} \int_0^\infty e^{-E/k_B T} \delta_{^1S_0}(E)\, dE. \tag{6.23}$$

This gives the EoS in the low-density limit

$$n_b^{(0)}(T, \mu_p, \mu_n) = n_{\text{free}}^{(0)}(T, \mu_p, \mu_n) + n_{\text{corr}}^{(0)}(T, \mu_p, \mu_n), \tag{6.24}$$

with

$$n_{\text{free}}^{(0)}(T, \mu_p, \mu_n) = \frac{2}{\Lambda^3} \left[e^{\mu_p/k_B T} + e^{\mu_n/k_B T} \right], \tag{6.25}$$

$$n_{\text{corr}}^{(0)}(T, \mu_p, \mu_n) = \frac{2^{5/2}}{\Lambda^3} \left\{ - \frac{e^{2\mu_p/k_B T}}{8} - \frac{e^{2\mu_n/k_B T}}{8} + 3 \left[e^{(\mu_p+\mu_n-E_d^{(0)})/k_B T} - 1 \right] + \sum_\tau \frac{g_\tau}{\pi k_B T} \int e^{(\mu_\tau - E)/k_B T} \delta_\tau(E)\, dE \right\}. \tag{6.26}$$

which is the rigorous result for the second virial coefficient and has to be considered as a low-density benchmark for the nuclear matter EoS. It should be mentioned that the Brueckner Hartree–Fock and any other quasiparticle or mean field approach cannot give the correct low-density limit for the deviation from the ideal nucleon gas. In addition to the free quasiparticle gas, we have to include the correlated part of the density, in particular the formation of bound states.

An extensive discussion on the ordinary Beth–Uhlenbeck formula and related EoS has been given recently by Horowitz and Schwenk [14]. Input quantities are the deuteron binding energy and the scattering phase shifts, which are experimentally accessible, without introducing a potential and solving the T-matrix equation. The generalized Beth–Uhlenbeck formula (6.19) derived in [6], see also [4], contains in-medium modifications of the single-particle and two-particle states in matter. It includes the Mott effect describing the dissolution of bound states at increasing nuclear matter density as discussed below.

Furthermore, it describes degeneration and, in particular, the occurrence of superfluidity at low temperatures. To obtain explicit results, we introduce a nucleon–nucleon interaction potential and solve the T-matrix equation in Sect. 6.5.

A phenomenological extension of the virial expansion has been proposed by Horowitz and Schwenk [14] who considered not only the nucleon–nucleon phase shifts, but also included further bound states such as α particles as well as Triton's and Helium 3 [15] and the scattering phase shifts between the different clusters to obtain the corresponding higher order virial coefficients. Although this cluster–virial expansion is not derived in a systematic way and double counting has to be avoided, it reflects the chemical picture and may be of use in the region where special clusters are dominating, such as α matter. A systematic treatment of all clusters where the interaction between the clusters is taken in mean-field approximation will be given in the following section.

6.4 Cluster Mean-Field Approximation

The Hartree–Fock approximation is the simplest approximation that takes the interaction into account in first order. Thus, any correlation is neglected, and we can introduce the concept of a mean field. Within the chemical picture where bound states are considered as new species, to be treated on the same level as free particles, a mean-field approach can be formulated by specifying the Feynman diagrams that are taken into account when treating A-particle cluster propagation [5]. The corresponding A-particle cluster self-energy is treated to first order in the interaction with the single particles as well as with the B-particle cluster states in the medium, but with full antisymmetrization between both clusters A and B. We use the notation $\{A, \nu, P\}$ for the particle number, internal quantum number (including proton number Z), and center of mass momentum for the cluster under consideration and $\{B, \bar{\nu}, \bar{P}\}$ for a cluster of the surrounding medium.

For the A-particle problem, the *effective wave equation* reads

$$\begin{aligned}
&[E(1) + \ldots E(A) - E_{A\nu P}]\psi_{A\nu P}(1 \ldots A) \\
&+ \sum_{1' \ldots A'} \sum_{i<j}^{A} V_{ij}^{A}(1 \ldots A, 1' \ldots A')\psi_{A\nu P}(1' \ldots A') \\
&+ \sum_{1' \ldots A'} V_{\mathrm{nm}}^{A,\mathrm{mf}}(1 \ldots A, 1' \ldots A')\psi_{A\nu P}(1' \ldots A') = 0, \qquad (6.27)
\end{aligned}$$

with $V_{ij}^{A}(1 \ldots A, 1' \ldots A') = V(12, 1'2')\delta_{33'} \ldots \delta_{AA'}$. The *effective potential* $V_{\mathrm{nm}}^{A,\mathrm{mf}}(1 \ldots A, 1' \ldots A')$ describes the influence of the nuclear medium on the cluster bound states and has the form

$$V_{\mathrm{nm}}^{A,\mathrm{mf}}(1 \ldots A, 1' \ldots A') = \sum_{i} \Delta(i)\delta_{11'} \ldots \delta_{AA'} + {\sum_{i,j}}' \Delta V_{ij}^{A}(1 \ldots A, 1' \ldots A')\,, \qquad (6.28)$$

with

$$\Delta(1) = \sum_2 (V(12,12)_{\rm ex}\tilde{f}(2) - \sum_{B=2}^{\infty} \sum_{\bar{\nu}\bar{P}} \sum_{2\ldots B} \sum_{1'\ldots B'} f_B(E_{B\bar{\nu}\bar{P}})$$

$$\times \sum_{i<j}^{m} V_{ij}^B(1\ldots B, 1'\ldots B')\psi_{B\bar{\nu}\bar{P}}(1\ldots B)\psi^*_{B\bar{\nu}\bar{P}}(1'\ldots B')\,, \qquad (6.29)$$

$$\Delta V_{12}^A(1\ldots A, 1'\ldots A') = -\Bigg\{\frac{1}{2}(\tilde{f}(1) + \tilde{f}(1'))V(12, 1'2') +$$

$$+ \sum_{B=2}^{\infty} \sum_{\bar{\nu}\bar{P}} \sum_{\bar{2}\ldots\bar{B}} \sum_{\bar{2}'\ldots\bar{B}'} f_B(E_{B\bar{\nu}\bar{P}}) \sum_{j}^{B} V_{1j}^B(1\bar{2}'\ldots\bar{B}', 1'\bar{2}\ldots\bar{B})$$

$$\times\, \psi^*_{B\bar{\nu}\bar{P}}(2\bar{2}\ldots\bar{B})\psi_{B\bar{\nu}\bar{P}}(2'\bar{2}'\ldots\bar{B}')\Bigg\}\delta_{33'}\ldots\delta_{AA'}\,, \qquad (6.30)$$

$$\tilde{f}(1) = f_1(1) + \sum_{B=2}^{\infty} \sum_{\bar{\nu}\bar{P}} \sum_{2\ldots B} f_B(E_{B\bar{\nu}\bar{P}})|\psi_{B\bar{\nu}\bar{P}}(1\ldots B)|^2\,, \qquad (6.31)$$

where (see (6.6))

$$f_A(E,\tau) = f_{A,Z_\tau}(E) = \left[\exp\beta(E-\mu_\tau) - (-1)^A\right]^{-1}\,. \qquad (6.32)$$

We note that within the mean-field approximation, the effective potential $V_{\rm nm}^{A,\rm mf}$ remains energy independent, that is, instantaneous. The quantity $\tilde{f}(1)$ describes the *effective occupation* of state 1 due to free and bound states, while exchange is included by the additional terms in ΔV_{12}^A and $\Delta(1)$, thus accounting for antisymmetrization.

The cluster mean-field may be viewed as a generalization of the ordinary mean field, where in addition to the mean field produced by the single-particle states, the mean field produced by clusters (bound states) is also taken into account. The modification of bound-state energies as well as wave functions can be evaluated in this approximation. We obtain an optimized set of states, which may be of use in evaluating self-energies and spectral functions in a consistent manner, as a prerequisite to evaluating correlation functions and thermodynamic relations. Note that the cluster mean-field approximation, which considers the interaction between the cluster A and B only in the lowest order of interaction, can be extended to higher orders of interaction. Then, we have to consider the complex of $(A+B)$ nucleons as a new few-nucleon system and have to calculate the quantum states solving the wave equation for the $(A+B)$ nucleon system.

Of course, the self-consistent solution of the cluster in a clustered medium is a rather involved problem, which has not been solved until now. In particular, the composition of the medium has to be determined, and for this we need the energy shift of the different components (clusters of B nucleons) in the medium. The energy shift of the respective cluster (B), in turn, has to be evaluated solving the effective wave equation for the B-nucleon problem.

Referring back to the approximation sketched above in which the medium is considered as uncorrelated, only the medium terms with $\tilde{f}(1)$ survive. All the higher-cluster distribution functions f_B are neglected, but $\tilde{f}(1)$ now denotes the Fermi distribution function for which the effective chemical potential $\tilde{\mu}_1$ is determined as reproducing the total nucleon density, $\Omega^{-1}\sum_{p_1} \tilde{f}_1(E^{\rm qu}(1)) = n_{\tau_1}$. This normalization reflects that all nucleons, bound or free, act as fermions and occupy phase space, resulting in Pauli blocking and self-energy shifts. Clearly, this approximation is most appropriate in that regions of the phase diagram where the contribution of clusters to the total density is small, that is, at high temperatures and low densities. This approximation is also suitable at densities above the Mott density, where the correlations have been destroyed due to the Pauli blocking.

We discuss the cluster decomposition of the self-energy once more so that the A-particle contribution follows from the A-particle T matrix. The A-particle T matrix obeys a Bethe–Salpeter equation, see (6.7),

$$\begin{aligned} G_A(1\dots A,1'\dots A',z_A) &= G_A^{(0)}(1\dots A,z_A)\delta_{11'}\dots\delta_{AA'} \\ &+ \sum_{1''\dots A''} G_A^{(0)}(1\dots A,z_A)V_A(1\dots A,1''\dots A'')G_A(1''\dots A'',1'\dots A',z_A), \end{aligned} \tag{6.33}$$

where $V_A(1\dots A,1'\dots A') = \sum_{i<j} V(ij,i'j')\prod_{k\neq i,j}\delta_{kk'}$ is the interaction within the A-particle cluster. z_A denotes a fermionic or bosonic Matsubara frequency. The free A-particle Green function follows as (see (6.5))

$$G_A^{(0)}(1\dots A,z_A) = \frac{[1-f_1(1)]\dots[1-f_1(A)]-(-1)^A f_1(1)\dots f_1(A)}{z_A - E^{\rm qu}(1) - \dots - E^{\rm qu}(A)}. \tag{6.34}$$

The solution of the Bethe–Salpeter equation follows as (6.8)

$$G_A(1\dots A,1'\dots A',z_A) = \sum_{\nu P} \psi_{A\nu P}(1\dots A)\frac{1}{z_A - E_{A\nu P}}\psi^*_{A\nu P}(1'\dots A'). \tag{6.35}$$

The summation over the internal quantum states ν includes besides the bound states also the scattering states. The A-particle wave function and the corresponding eigenvalues follow from solving the in-medium Schrödinger equation

$$\begin{aligned} &[E^{\rm qu}(1)+\dots+E^{\rm qu}(A)-E_{A\nu P}]\psi_{A\nu P}(1\dots A) \\ &+ \sum_{1'\dots A'}\sum_{i<j}[1-\tilde{f}_1(i)-\tilde{f}_1(j)]V(ij,i'j')\prod_{k\neq i,j}\delta_{kk'}\psi_{A\nu P}(1'\dots A') = 0\,. \end{aligned} \tag{6.36}$$

The main structures are given in this equation: we have dressed single-particle propagators given by quasiparticle energies. This equation contains the effects of the medium in the quasiparticle shift as well as in the Pauli blocking terms. The approximation of an uncorrelated medium leads to the effective occupation numbers $\tilde{f}_1(1) = [\exp((E^{\mathrm{qu}}(1) - \tilde{\mu}_1)/k_{\mathrm{B}}T) + 1]^{-1}$, where $\tilde{\mu}_1$ is determined by the normalization condition $\Omega^{-1}\sum_{p_1} \tilde{f}_1(1) = n_{\tau_1}$. A more detailed description of the medium containing also the correlations is given by the cluster mean-field approximation, where besides the single particle distribution function f_1 also higher cluster distribution functions f_A occur.

Obviously, the bound state wave functions and energy eigenvalues as well as the scattering phase shifts become dependent on temperature and density. Two effects have to be considered, the quasiparticle energy shift and the Pauli blocking.

6.5 Nucleon–Nucleon Interaction

In contrast to the second virial coefficient, which can be expressed in terms of observables such as the deuteron binding energy and the scattering phase shifts in the spin singlet or in the spin triplet channels, the evaluation of medium modifications of the single and two-particle states can be performed only if the interaction potential is known. As long as the nucleon–nucleon interaction reproduces the deuteron binding energy and the scattering phase shifts in the different channels, the low-density expansion coincides with the ordinary Beth–Uhlenbeck formula for the second virial coefficient.

The full treatment of the empirical data to construct separable potentials has been performed by Plessas et al. [13], who presented a rank 4 separable interaction based on the Paris (PEST) or Bonn (BEST) potential. This potential has been used in calculating correlations and bound state formation in nuclear matter, see [6, 8]. We give here some simpler rank 1 separable forms [16] based on the following two-particle properties: The deuteron binding energy is $E_{\mathrm{d}} = 2.224573\,\mathrm{MeV}$. In the spin triplet state, $\tau = {}^3S_1$, we have besides the bound state properties the observed scattering phase shifts, which are characterized by the scattering length $a_{^3S_1} = 5.396\,\mathrm{fm}$ and the effective range $r_{^3S_1} = 1.726\,\mathrm{fm}$. In the spin singlet state, $\tau = {}^1S_0$, we have no bound state. From the observed scattering phase shifts we have the scattering length $a_{^1S_0} = -23.678\,\mathrm{fm}$ and the effective range $r_{^1S_0} = 2.729\,\mathrm{fm}$. Physical constants are taken as $\hbar c = 197.327\,\mathrm{MeV\,fm}$, the isospin averaged mass $mc^2 = 939.1735\,\mathrm{MeV}$, so that $\hbar^2/m = 41.4598\,\mathrm{MeV\,fm}^2$.

For the two-nucleon system, we take a Gaussian interaction of the form

$$\begin{aligned} V_\tau(p_1, p_2; p_1', p_2') &= -\lambda_\tau w(p) w(p') \delta_{p_1+p_2, p_1'+p_2'} \\ &= -\lambda_\tau e^{-p^2/\beta_\tau^2} e^{-p'^2/\beta_\tau^2} \delta_{p_1+p_2, p_1'+p_2'}, \end{aligned} \tag{6.37}$$

with $p = (p_2 - p_1)/2, p' = (p_2' - p_1')/2$. Better approximations are given by coupled waves and higher rank interactions, see [13].

The advantage of a separable interaction is that the ladder-T matrix (6.12) can be solved. We give the solution for uncoupled waves in the channel $\tau = {}^3S_1$ (isospin singlet, spin triplet) and $\tau = {}^1S_0$ (isospin triplet, spin singlet) including the Pauli blocking factors, starting with

$$\mathrm{T}_\tau(12,1'2',z) = V_\tau(12,1'2') + \sum_{1"2"} V_\tau(12,1"2") \frac{1-f_1(1")-f_1(2")}{z-E^{\mathrm{qu}}(1")-E^{\mathrm{qu}}(2")} T_\tau(1"2",1'2',z). \tag{6.38}$$

For the separable interaction (6.37), we take the ansatz $\mathrm{T}_\tau(12,1'2',z) = w(p)w(p')t_\tau(P,z)\delta_{p_1+p_2,P}\delta_{p'_1+p'_2,P}$ and find

$$t_\tau(P,z) = -\lambda_\tau - \lambda_\tau \sum_p w^2(p) \frac{1-f_1(\boldsymbol{p}+\boldsymbol{P}/2)-f_1(\boldsymbol{p}-\boldsymbol{P}/2)}{z-E^{\mathrm{qu}}(\boldsymbol{p}+\boldsymbol{P}/2)-E^{\mathrm{qu}}(\boldsymbol{p}-\boldsymbol{P}/2)} t_\tau(P,z), \tag{6.39}$$

with the solution

$$t_\tau(P,z) = -\lambda_\tau \left[1+\lambda_\tau \sum_p w^2(p) \frac{1-f_1(\boldsymbol{p}+\boldsymbol{P}/2)-f_1(\boldsymbol{p}-\boldsymbol{P}/2)}{z-E^{\mathrm{qu}}(\boldsymbol{p}+\boldsymbol{P}/2)-E^{\mathrm{qu}}(\boldsymbol{p}-\boldsymbol{P}/2)}\right]^{-1}. \tag{6.40}$$

The binding energy is obtained as the value of z where the denominator of $t_\tau(P,z)$ (see (6.40)) becomes equal to zero. The scattering phase shift in the channel τ is given by

$$\delta_\tau(P,E) = \arctan\left[\frac{\mathrm{Im}\, t_\tau(P,E)}{\mathrm{Re}\, t_\tau(P,E)}\right]. \tag{6.41}$$

The scattering length a_τ and the effective range $r_{0,\tau}$ follow from $k\cot\delta_\tau = -1/a_\tau + r_{0,\tau}k^2/2 - P_\tau r_{0,\tau}^3 k^4 + \cdots$.

We evaluate T_2^L for the Gaussian form-factor

$$\mathrm{T}_\tau(k_1k_2,k'_1k'_2,z) = -\lambda_\tau\, \mathrm{e}^{(k_1-k_2)^2/(4\beta^2)} \mathrm{e}^{(k'_1-k'_2)^2/(4\beta^2)} t_\tau(P,T,n_p,n_n,z), \tag{6.42}$$

with

$$\begin{aligned}
\frac{1}{t_\tau(P,T,n_p,n_n,z)} &= 1 \\
&- \frac{\lambda_\tau}{2\pi^2}\frac{m^*}{\hbar^2}\frac{\beta}{2^{3/2}}\left(\pi^{1/2} - \pi\left(-\frac{2m^*\bar{z}}{\hbar^2\beta^2}\right)^{1/2} \mathrm{e}^{-\frac{2m^*\bar{z}}{\hbar^2\beta^2}}\, \mathrm{Erfc}\left[\left(-\frac{2m^*\bar{z}}{\hbar^2\beta^2}\right)^{1/2}\right]\right) \\
&+ \frac{n_p+n_n}{2}\left(\frac{2\pi\hbar^2}{m^*k_{\mathrm{B}}T}\right)^{3/2} \mathrm{e}^{-\hbar^2P^2/8m^*k_{\mathrm{B}}T}\, \frac{\lambda_\tau}{2\pi^2}\frac{m^*}{\hbar^2}\left(\frac{2}{\beta^2}+\frac{\hbar^2}{2m^*k_{\mathrm{B}}T}\right)^{-1/2} \\
&\times \frac{1}{2}\left(\pi^{1/2} - \pi\left(-\frac{m^*\bar{z}}{\hbar^2}\left(\frac{2}{\beta^2}+\frac{\hbar^2}{2m^*k_{\mathrm{B}}T}\right)\right)^{1/2} \mathrm{e}^{-\frac{m^*\bar{z}}{\hbar^2}\left(\frac{2}{\beta^2}+\frac{\hbar^2}{2m^*k_{\mathrm{B}}T}\right)}\right. \\
&\left.\mathrm{Erfc}\left[\left(-\frac{m^*\bar{z}}{\hbar^2}\left(\frac{2}{\beta^2}+\frac{\hbar^2}{2m^*k_{\mathrm{B}}T}\right)\right)^{1/2}\right]\right).
\end{aligned} \tag{6.43}$$

Here, the Fermi distribution and the quasiparticle shifts in (6.40) are angle averaged, and the nondegenerate distribution $f_1(p) = n_{\tau_1} \Lambda^3 \exp[-p^2/2m^* k_\mathrm{B} T]$ is used, with $\bar{z} = z - 2v - P^2\hbar^2/(4m^*)$ reflecting the quasiparticle properties as described by the single nucleon shift v and the effective mass m^*. Dropping all density effects, a more simple form results as the solution of the isolated two-nucleon problem

$$t_\tau(E) = \left[1 - \frac{\lambda_\tau}{2\pi^2}\frac{m}{\hbar^2}\frac{\beta}{2^{3/2}}\left(\pi^{1/2} - \pi\left(-\frac{2mE}{\hbar^2\beta^2}\right)^{1/2} \mathrm{e}^{-\frac{2mE}{\hbar^2\beta^2}} \mathrm{Erfc}\left[\left(-\frac{2mE}{\hbar^2\beta^2}\right)^{1/2}\right]\right)\right]^{-1} . \tag{6.44}$$

Considering the solution of the isolated two-nucleon problem, we can determine the parameter of the potential (6.37). Similar to Yamaguchi [16], we use the deuteron bound state energy and the scattering length to fix the two parameters $\lambda_{^3S_1} = 1172.87\,\mathrm{MeV\,fm}^3$, $\beta_{^3S_1} = 1.58\,\mathrm{fm}^{-1}$ (spin triplet) (units: MeV, fm)

$$V_{^3S_1}(p, p') = -1172.58\,\mathrm{e}^{-(p^2+p'^2)/1.58^2} . \tag{6.45}$$

For the spin singlet state we adapt like Yamaguchi the same parameter $\beta_{^1S_0} = \beta_{^3S_1} = \beta$. From the fit to the scattering length follows $\lambda_{^1S_0} = 774.688\,\mathrm{MeV\,fm}^3$ (spin singlet)

$$V_{^1S_0}(p, p') = -774.688\,\mathrm{e}^{-(p^2+p'^2)/1.58^2} . \tag{6.46}$$

From the solution of this potential, we calculate in the spin triplet channel the nucleonic rms value $(\langle r^2\rangle_\mathrm{nucleon})^{1/2} = 1.928\,\mathrm{fm}$ for the deuteron (see Sect. 6.8) and the effective range $r_{0,^3S_1} = 1.729\,\mathrm{fm}$ for the scattering phase shifts, whereas for the spin triplet channel $r_{0,^1S_0} = 2.0979\,\mathrm{fm}$ is calculated. These values give a good parametrization of the low-energy two-nucleon properties. Better coincidence is possible using more sophisticated potentials [13] where partial waves for uncoupled channels (singlet, $^1S_0$) and coupled channels (triplet, $J = 1$, $^3S_1 -^3 D_1$), etc. are treated, and the set of adjustable parameters is extended. Different angular momenta with higher rank basis systems are considered.

We can also perform the fit to a more complex (rank 2) separable potential, reflecting attraction and repulsion, see also [16]. The ratio of both ranges of interaction can be related to the masses of σ and ω mesons. Using units MeV and fm for the spin triplet channel follows

$$V_{^3S_1}(p, p') = -1645.89\,\mathrm{e}^{-(p^2+p'^2)/1.749^2} + 445.843\,\mathrm{e}^{-(p^2+p'^2)/2.49^2} \tag{6.47}$$

and for the spin singlet channel

$$V_{^1S_0}(p, p') = -1309.436\,\mathrm{e}^{-(p^2+p'^2)/1.749^2} + 484.658\,\mathrm{e}^{-(p^2+p'^2)/2.49^2} . \tag{6.48}$$

In the following, we restrict us to the simple rank 1 potential (6.45) and (6.46) to show the effects we are interested in.

An alternative is a Lorentz-type form-factor, which, however, does not allow to separate the center-of-mass (c.o.m.) momentum in a simple way:

$$\langle k, K|V|k', K'\rangle = V(k,k')\,\delta_{K,K'}\,,$$
$$V_\tau(k,k') = \lambda_\tau w(k) w(k')\,, \qquad w(k) = \frac{1}{k^2+\beta^2}. \tag{6.49}$$

According to Yamaguchi [16] we have

$$(k^2-p^2)\psi(p) = -\lambda_\tau \frac{1}{p^2+\beta^2}\int \mathrm{d}^3p'\,\frac{1}{p'^2+\beta^2}\psi(p'), \tag{6.50}$$

with $\beta = 1.4488\,\mathrm{fm}^{-1}$, $\lambda_{^3S_1} = 8\pi\beta^3\frac{\hbar^2}{m}\left(1+\sqrt{-\frac{m}{\hbar^2\beta^2}(-E_\mathrm{D})}\right)^2 = 4263.05$ MeV fm^{-1}, and find

$$\bar\lambda_{^3S_1} = \pi^{-2}\beta(\alpha+\beta)^2,\; E_\mathrm{D} = \alpha^2/m, \tag{6.51}$$

with $1/a = (\beta/2)\left[1-\beta^3/(\pi^2\bar\lambda)\right]$ and $r_0 = (1/\beta)\left[1+2\beta^3/(\pi^2\bar\lambda)\right]$. A consistency test is $r_{0,^3S_1} = 2/k_\mathrm{D}\,[1-1/(a_{^3S_1}k_\mathrm{D})] = 1.726\,\mathrm{fm}$, with $k_\mathrm{D}^2\hbar^2/m = E_\mathrm{D} = 2.224573\,\mathrm{MeV}$ and $\alpha_{^3S_1} = 0.2319\,\mathrm{fm}^{-1}$, thus $\bar\lambda_{^3S_1} = \lambda_{^3S_1}m/(8\pi^3\hbar^2)$. For the isolated two-nucleon system, the integral in $t_\tau(P,E)$, (6.40), can be performed with the result

$$-\frac{\lambda_{^3S_1}}{2\pi^2}\frac{m}{\hbar^2}\int\frac{p^2\,\mathrm{d}p}{(p^2+\beta^2)^2}\frac{1}{(-Em/\hbar^2)+p^2} = \frac{\pi}{4\beta^3}\frac{-5.20911}{(1+\sqrt{-Em/\hbar^2\beta^2})^2}$$
$$= -\frac{1.34533}{(1+0.107196\sqrt{(-E)})^2}\,. \tag{6.52}$$

For the singlet channel we find $\lambda_{^1S_0} = 2550.03\,\mathrm{MeV\,fm}^{-1}$.

In conclusion, using higher rank separable interactions and coupled channels, any potential can be represented with arbitrary accuracy, see [13]. For exploratory calculations, we restrict us here to a simple Gaussian rank 1 separable interaction (6.45) and (6.46).

6.6 Quasiparticle Approximation and the EoS at High Densities

Starting from the noninteracting, ideal Fermi gas to describe nuclear matter, the simplest approximation which is of first order in the interaction potential is the Hartree–Fock approximation, where the single-particle energies are shifted by $\Delta^\mathrm{HF}(1) = \sum_2 V(12,12)_\mathrm{ex}[\exp((p_2^2/2m+\Delta^\mathrm{HF}(2)-\mu_2)/k_\mathrm{B}T)+1]^{-1}$. Using an appropriate nucleon–nucleon interaction potential, the EoS can be evaluated in Hartree–Fock approximation (6.17) with $E^\mathrm{qu}(1) = p_1^2/2m+\Delta^\mathrm{HF}(1)$. In a simple approximation, we can introduce the quasiparticle shift and effective mass to describe the dispersion relation at small momenta. However, this approximation is not sufficient, because it does not contain the relevant correlations. As already mentioned, the deviation of the EoS from the ideal behavior in the low-density limit is given by the virial expansion. In particular, bound state formation, which has to be taken into account to obtain the correct virial expansion at low densities, is not described in Hartree–Fock approximation.

At saturation and higher densities, correlations are suppressed due to Pauli blocking when the Fermi energy exceeds the interaction energy. Then, a quasiparticle concept can be used to derive the EoS of nuclear matter, taking into account only the singularity in the spectral function (6.3) for vanishing $\mathrm{Im}\,\Sigma$, and also in this case nuclear matter is not correctly described in Hartree–Fock approximation. The phenomenological properties derived from nuclear structure and excited nuclei as well as the expected properties of neutron star core [2] give benchmarks for a realistic nuclear matter EoS. The main shortcoming of the Hartree–Fock approximation is the strong repulsion at short nucleon distances, which demands the treatment of short-range correlations absent in a mean-field approach.

A microscopic calculation of the quasiparticle energies has been performed within the Brueckner–Hartree–Fock calculations, which also have been made for the Dirac theory [17]. The relation to the formalism presented above, in particular solving the T matrix for a separable interaction, is shown in [6]. These calculations give a realistic description also near the saturation density. At low densities, the formation of bound states is of importance, and a self-consistent solution of the quasiparticle energy may become impossible when the structure of the spectral function becomes complex.

Semiempirical expressions have been proposed to model the known properties of nuclear matter. Well known are the Skyrme parametrization of the quasiparticle energy shifts, see [4, 18], which are improved presently by the relativistic mean field approaches, see [12, 19]. These approaches are based on a field theory, which reflects essential symmetries of the fundamental microscopic theory, but combine results of the microscopic calculations such as Dirac–Brueckner Hartree–Fock [17] with empirical data to adjust parameters so that a simple mean-field approximation contains already the relevant many-particle effects. This way, a nuclear matter EoS is obtained, which is applicable also near and above saturation density. The main issue is an optimal quasiparticle description. We will not give here the details, but only mention that at present they are successfully used to calculate the nuclear structure. They are also of interest for the structure of compact objects such as neutron stars, supernovae explosions, and the early universe, see [2].

We conclude that an appropriate introduction of the quasiparticle energy will be of importance to get the correct description of nuclear matter at higher densities when the contribution of bound states disappear. The nuclear EoS then is given by the quasiparticle contribution (6.17). It is a challenge to an advanced nuclear matter EoS to merge from the correct low-density limit given by the virial expansion to the quasiparticle description at high densities.

The quasiparticle shift occurs also in the few-body wave equations (6.36). The properties of the bound states in a dense medium will be modified due to the Pauli blocking. In addition, the properties of bound states will be modified if the quasiparticle shift, which arises from the self-energy, is considered, as shown in the following. Improving former approaches [4,20] where the Skyrme parametrization has been used, we apply the parametrization according to the relativistic mean field approaches, see [12,19], which give expressions for the scalar as well as for the vector potential energy.

6.7 Medium Modifications of Two-Particle Correlations

With increasing density of nuclear matter, medium modifications of single-particle states as well as of few-nucleon states become of importance. The self-energy of an A-particle cluster can in principle be deduced from contributions describing the single-particle self-energies as well as medium modifications of the interaction and the vertexes. A guiding principle in incorporating medium effects is the construction of *consistent* ("conserving") approximations, which treat medium corrections in the self-energy and in the interaction vertex at the same level of accuracy. This can be achieved in a systematic way using the Green functions formalism. At the mean-field level, we have only the Hartree–Fock self-energy $\Delta^{\rm HF} = \sum_2 V(12,12)_{\rm ex} \tilde{f}_1(2)$ together with the Pauli blocking factors, which modify the interaction from $V(12,1'2')$ to $V(12,1'2')[1-\tilde{f}_1(1)-\tilde{f}_1(2)]$. More advanced, within the cluster-mean field approximation, the few-body wave equation (6.36) was obtained, which consistently describes the modification of the few-body system due to the embedding correlated nuclear matter.

In the case of the two-nucleon system ($A=2$), the resulting effective wave equation that includes the mean-field corrections reads

$$
\begin{aligned}
&[E^{\rm qu}(1) + E^{\rm qu}(2) - E_{2,\nu,P}]\, \psi_{2,\nu,P}(12) \\
&+ \sum_{1'2'} [1 - \tilde{f}_1(1) - \tilde{f}_1(2)]\; V(12,1'2') \psi_{2,\nu,P}(1'2') = 0. \qquad (6.53)
\end{aligned}
$$

This *effective wave equation* describes bound states as well as scattering states. As also shown below in Sect. 6.11, we also mention that the Gor'kov equation describing the transition to superfluidity is reproduced from (6.53) when the binding energy $E_{2,\nu,P=0} = E_{d,P=0}$ coincides with $\mu_p + \mu_n$.

From the solution of the in-medium two-particle Schrödinger equation (6.53) or the corresponding T matrix, the scattering and possibly bound states are obtained. Because of the self-energy shifts and the Pauli blocking, the binding energy of the deuteron $E_d(P,T,\mu_p,\mu_n)$ as well as the scattering phase shifts $\delta_\tau(E,P,T,\mu_p,\mu_n)$ in the isospin singlet or triplet channel τ, respectively, will depend on the temperature and the chemical potentials. For a separable interaction $V(12,1'2')$ like the PEST4 potential [13], an analytical solution of (6.53) can be found in the low-density limit, and the results for the shift of the binding energy and the medium modification of the scattering phase shifts are discussed extensively, see [6,8]. We discuss the medium shift of the binding energy in perturbation theory.

We take the quasi-particle energies that are described by an effective mass and a self-energy shift and solve the Schrödinger equation for a separable potential. Separating the center of mass motion, with energy $p^2/2\,(m_p^* + m_n^*)$ from the relative motion, with reduced mass $m_p^* m_n^*/(m_p^* + m_n^*)$, we find the binding energy $E_d^{\mathrm{qu}}(P) = E_d + \Delta E_d^{\mathrm{SE}} + \Delta E_d^{\mathrm{Pauli}}$, where E_d is density dependent due to the effective masses. The corresponding wave function is used to evaluate the Pauli blocking term

$$\Delta E_d^{\mathrm{Pauli}} = \sum_{1\,2\,1'\,2'} \psi_{d,P}(1\,2)\, V(1\,2, 1'\,2')\, [\tilde{f}_1(1) + \tilde{f}_1(2)]\, \psi_{d,P}^*(1'\,2') \tag{6.54}$$

in first order perturbation theory. The self-energy shift ΔE_d^{SE} is simply the sum of the quasi-particle self-energy shift $v(1)$ of the proton and neutron, (6.15) at $p_1 = 0$, as discussed in the previous section.

In Fig. 6.1, we show the shift of the binding energy of the deuteron (d) with zero c.o.m. momentum in symmetric nuclear matter as a function of density for temperature $T = 10\,\mathrm{MeV}$. The shift of the other light clusters ($t/h, \alpha$), also shown in Fig. 6.1, will be discussed in the following section.

It is found that the cluster binding energy decreases with increasing density. Finally, at the *Mott density* $n_{A,\nu,P}^{\mathrm{Mott}}(T,\alpha)$, depending on the temperature T

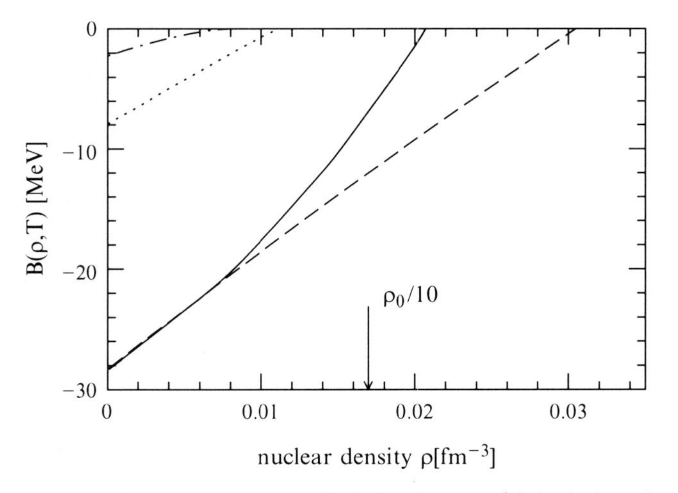

Fig. 6.1. Shift of binding energy B of the light clusters (d *dash dotted*, t/h *dotted*, and α *dashed*: perturbation theory, *full line*: nonperturbative Faddeev–Yakubovski equation) in symmetric nuclear matter as a function of density $\rho = n_{\mathrm{B}}$ for given temperature $T = 10\,\mathrm{MeV}$ [7]

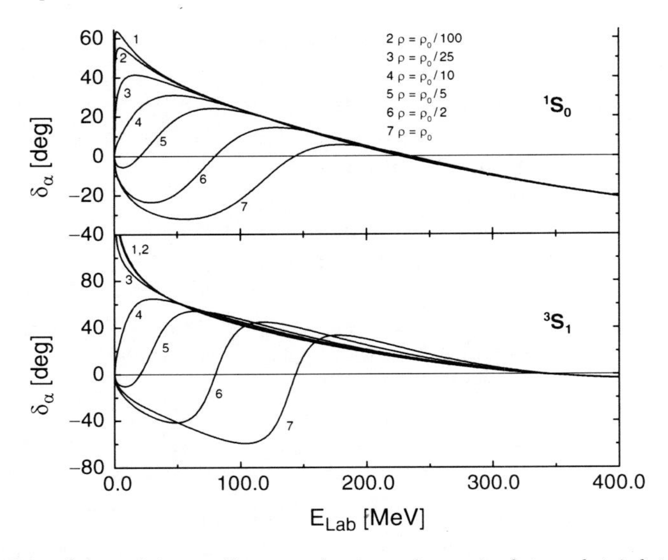

Fig. 6.2. Free (1) and in-medium nucleon–nucleon singlet and triplet scattering phase shifts in the S channel for different baryon densities as function of the energy $E_{\rm Lab}$, $T = 10\,{\rm MeV}$

and the asymmetry parameter α, the bound state is dissolved. The clusters are not present at higher densities, having been merged into the nucleonic quasi-particle liquid. For a given cluster type, characterized by $\{A,\nu\} = \{A,Z,n\}$, we can also introduce the Mott momentum $P^{\rm Mott}_{A,\nu}(n_p,n_n,T)$ depending on the ambient temperature T and nucleon densities n_p,n_n, such that bound states exist only for $P \geq P^{\rm Mott}_{A,\nu}(n_p,n_n,T)$.

In-medium scattering phase shifts are obtained from the two-particle T matrix, see (6.41). At given temperature $T = 10\,{\rm MeV}$ and different baryonic densities, the singlet and triplet phase shifts are shown in Fig. 6.2 for symmetric matter.

As a consequence, the virial expansion of the EoS (generalized Beth–Uhlenbeck formula (6.17) and (6.18)) for the total baryon density $n_{\rm B}$

$$n_{\rm B}(T,\mu_p,\mu_n) = n_{\rm free}(T,\mu_p,\mu_n) + n_{\rm corr}(T,\mu_p,\mu_n) \tag{6.55}$$

constitutes of the single-particle contributions $n_{\rm free} = n_p^{\rm free} + n_n^{\rm free}$, where $n_\tau^{\rm free}(T,\mu_\tau) = 2/(2\pi\hbar)^3 \int {\rm d}^3p f_{1,\tau}(E^{\rm qu}_\tau(p))$ describes the free quasi-particle contributions of protons ($\tau = p$) or neutrons ($\tau = n$), respectively, and the two-particle contributions $n_{\rm corr} = n_2^{\rm bound} + n_2^{\rm scat}$ containing the contribution of deuterons (spin factor 3)

$$n_2^{\rm bound}(T,\mu_p,\mu_n) = 3 \int_{P>P_d^{\rm Mott}} \frac{{\rm d}^3P}{(2\pi\hbar)^3}\, f_{2,d}(E^{\rm qu}_d(P)), \tag{6.56}$$

with $f_{2,d}(E_d^{\rm qu}(P)) = [{\rm e}^{(E_d^{\rm qu}(P;T,\mu_p,\mu_n)-\mu_p-\mu_n)/k_{\rm B}T} - 1]^{-1}$, and scattering states of the isospin singlet and triplet channel τ (degeneration factor g_τ)

$$n_2^{\rm scat}(T,\mu_p,\mu_n) = \sum_\tau g_\tau \int \frac{{\rm d}^3P}{(2\pi\hbar)^3} \int_0^\infty \frac{{\rm d}E}{2\pi}\, f_{2,\tau}(\Delta E_d^{\rm SE}(P) + E)$$
$$\times \sin^2 \delta_\tau(E,P;T,\mu_p,\mu_n) \frac{\rm d}{{\rm d}E} \delta_\tau(E,P;T,\mu_p,\mu_n), \tag{6.57}$$

$\Delta E_d^{\rm SE}(P;T,\mu_p,\mu_n)$ is the shift of the continuum edge (self-energies at momentum $P/2$).

The EoS (6.55) shows some interesting features: (1) In the low-density limit, a mass action law is obtained describing an ideal mixture of free nucleons and deuterons. We stress that a quasi-particle picture is not able to reproduce this important limiting case correctly. (2) With increasing density, the single-particle properties as well as the two-particle properties are simultaneously modified by the medium. In particular, the bound states are dissolved at high densities. (3) There is also a contribution from scattering states to the two-particle density. As a consequence of the Levinson theorem, the contribution of the disappearing bound states is replaced by a contribution from the scattering states (resonances) at the Mott density so that the total two-particle density n_2 behaves smoothly. (4) Because of the pole of the Bose distribution function at low temperatures, pairing can occur in n_2. A smooth transition from BEC of deuterons at low densities to Cooper pairing at high densities is observed [8].

Calculations of the composition ($n_2/n_{\rm B}$) of symmetric nuclear matter ($n_p = n_n$, no Coulomb interaction) are shown in Fig. 6.3 [8]. At low densities,

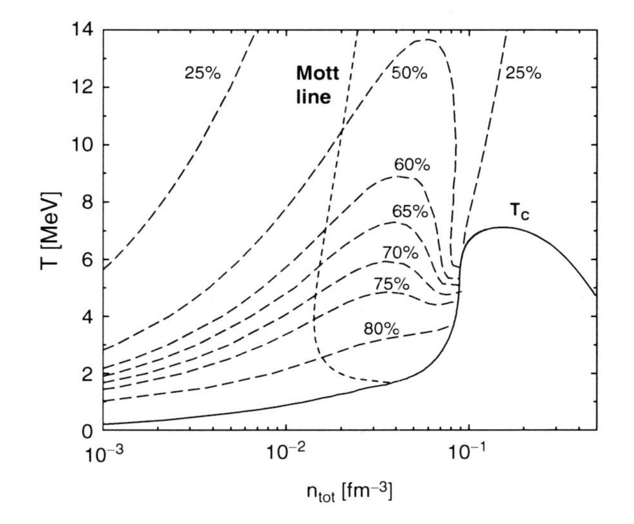

Fig. 6.3. Fraction of correlated density for symmetric nuclear matter as function of temperature T and total baryon density $n_{\rm tot} = n_{\rm B}$. Only two-particle correlations are taken into account

the contribution of bound states becomes dominant at low temperatures. At fixed temperature, the contribution of the correlated density n_2 is first increasing with increase in density according to the mass action law, but above the Mott line it is sharply decreasing, so that near nuclear matter saturation density ($n_\mathrm{B} = n_0 = 0.17\,\mathrm{fm}^{-3}$) the contribution of the correlated density almost vanishes. Also, the critical temperature T_c for the pairing transition is shown, which will be discussed in Sect. 6.11.

In the low-density limit, perturbation theory gives $E_d^\mathrm{qu}(P) = E_d^0 + P^2/2M_d + \Delta E_d^\mathrm{SE}(P) + \Delta E_d^\mathrm{Pauli}(P)$ with $E_d^0 = -2.22\,\mathrm{MeV}$, and

$$\begin{aligned}\Delta E_d^\mathrm{Pauli} &= \sum_{12,1'2'} \psi_{d,P}(12)[\tilde{f}_1(1) + \tilde{f}_1(2)]V(12,1'2')\psi_{d,P}(1'2') \\ &\approx \psi_{d,P}^2(0)(n_p + n_n)(E_d^0 - E_d^\mathrm{kin}),\end{aligned} \tag{6.58}$$

where E_d^kin denotes the mean kinetic energy of the nucleons in the unperturbed deuteron. To reproduce the behavior shown in Fig. 6.3, in [21] the following parametrization

$$\Delta E_d \approx 340\ (n_p + n_n)\ \mathrm{MeV\,fm}^3 + 13000\ (n_p^2 + n_n^2)\ \mathrm{MeV\,fm}^6 \tag{6.59}$$

was proposed for $T = 10\,\mathrm{MeV}$. The result of this calculation can be compared with the evaluation of the correlated density shown in [6]. Two particle correlations are suppressed for densities higher than the Mott density of about $0.001\,\mathrm{fm}^{-3}$, but will be present up to densities of the order of nuclear matter saturation density n_0. A more general expression for arbitrary temperatures is given in the following section.

6.8 Medium Modification of Cluster Properties

In the low-density limit, the most important effect of interaction with respect to the nuclear matter EoS is the formation of bound states characterized by the proton content Z and the neutron content $N = A - Z$. We restrict us to light clusters only with $A \leq 4$ as the α particle is strongly bound. Heavier nuclei can be treated in a similar way [4, 20]. Large strongly bound clusters such as iron, being of importance in considering the outer crust of neutron stars, for example, can be described using other concepts such as phase instability and the formation of a liquid droplet in nuclear matter, see, for example [12, 18].

After discussing the two-nucleon system in the previous section, we consider higher clusters. In-medium wave equations similar to (6.53) have been derived from the Green function approach for the case $A = 3$ and $A = 4$, (6.36) describing triton/helion (^{3}He) nuclei as well as α-particles embedded in nuclear matter. The effective wave equation contains the quasiparticle self-energy shift of the single-particle energies as well as the Pauli blocking of the interaction. We give the effective wave equation for $A = 4$,

$$0 = [E^{\rm qu}(1) + E^{\rm qu}(2) + E^{\rm qu}(3) + E^{\rm qu}(4) - E_{4,\nu,P}]\, \psi_{4,\nu,P}(12)$$
$$+ \sum_{i<j}^{4} \sum_{1'2'3'4'} [1 - \tilde{f}_1(i) - \tilde{f}_1(j)] V(ij, i'j') \prod_{k \neq i,j} \delta_{k,k'} \psi_{4,\nu,P}(1'2'3'4'). \quad (6.60)$$

A similar equation is obtained for $A = 3$.

We first discuss the low-density limit where any medium effects are omitted, that is, $\tilde{f}_1(i) \ll 1$, $E^{\rm qu}(1) = p_1^2/2m$. In contrast to the two-particle problem, where rigorous solutions are available, few-body techniques have to be applied to solve the three-particle or the four-particle problem. Here we restrict us to variational approaches and perturbation theory. We have to parameterize the nucleon–nucleon interaction in the corresponding channels that reproduces the empirical data. Considering t, h, α, we account this way for Coulomb repulsion and three-nucleon interaction what is well-known when introducing density-dependent nucleon–nucleon interactions.

In case of the two-nucleon system, the interaction is parametrized in the different channels, using empirical data. Adjusting the interaction potentials to two-particle scattering data and bound state properties as given in the previous section, the thermodynamic properties, which are directly related to the empirical data (mass action law and composition, second virial coefficient), are correctly reproduced.

Now we are considering higher clusters. Properties of light clusters are given in Table 6.1, see [22]. We used $\langle r^2 \rangle_{\rm nucleon} = \langle r^2 \rangle_{\rm electrom.} - \langle r^2 \rangle_{\rm proton}$. To describe the nucleon wave function of the light clusters, a Gaussian form-factor is assumed, as it allows for the separation of the c.o.m. motion in a simple way. We restrict us here to variational approaches in solving the few-nucleon problem. The potential appropriate to reproduce two-particle properties has already been given in the previous section. Parameters for the Gaussian-type nucleon–nucleon interaction (6.37) in the other cluster channels can be given, reproducing the measured binding energies and nucleonic rms values. Using a variational approach with Gaussian product wave functions,

$$\psi_{t/h}(p_1, p_2, p_3) = \text{const } \mathrm{e}^{-p_1^2/b^2 - p_2^2/b^2 - p_3^2/b^2}\, \delta_{p_1+p_2+p_3,0}, \quad (6.61)$$

Table 6.1. Properties of light nuclei

	Binding energy	Mass	Spin	rms-radius (electrom.)	rms-radius (nucleon)
n	0	939.565 MeV c^{-2}	1/2	0.34 fm	–
p (^{1}H)	0	938.783 MeV c^{-2}	1/2	0.87 fm	0 fm
d (^{2}H)	−2.225 MeV	1876.12 MeV c^{-2}	1	2.17 fm	1.99 fm
t (^{3}H)	−8.482 MeV	2809.43 MeV c^{-2}	1/2	1.70 fm	1.46 fm
h (^{3}He)	−7.718 MeV	2809.41 MeV c^{-2}	1/2	1.87 fm	1.66 fm
α (^{4}He)	−28.30 MeV	3728.40 MeV c^{-2}	0	1.63 fm	1.38 fm

or

$$\psi_\alpha(p_1, p_2, p_3, p_4) = \text{const } \mathrm{e}^{-p_1^2/b^2 - p_2^2/b^2 - p_3^2/b^2 - p_4^2/b^2} \delta_{p_1+p_2+p_3+p_4,0}, \qquad (6.62)$$

respectively, we have for the three-nucleon bound state (triton/helion)

$$\langle r^2 \rangle_{\text{nucleon}} = 2/b^2 \qquad (6.63)$$

so that $b_t = 0.968\,\text{fm}^{-1}$ and $b_h = 0.968\,\text{fm}^{-1}$. Determining the minimum of

$$E_3 = \frac{\hbar^2}{m}\frac{3}{4}b^2 - 3\lambda\frac{\beta^6 b^3}{\pi^{3/2}(b^2+\beta^2)^3}, \qquad (6.64)$$

we find $\lambda_t = 1416\,\text{MeV fm}^3$ and $\beta_t = 1.21\,\text{fm}^{-1}$, or $\lambda_h = 1735\,\text{MeV fm}^3$, $\beta_h = 1.04\,\text{fm}^{-1}$, respectively.

For the α-particle we have $\langle r^2 \rangle_{\text{nucleon}} = 9/(4b^2)$ so that $b_\alpha = 1.087\,\text{fm}^{-1}$. Determining the minimum of

$$E_4 = \frac{\hbar^2}{m}\frac{9}{8}b^2 - 6\lambda\frac{\beta^6 b^3}{\pi^{3/2}(b^2+\beta^2)^3}, \qquad (6.65)$$

we find $\lambda_\alpha = 1295\,\text{MeV fm}^3$, $\beta_\alpha = 1.231\,\text{fm}^{-1}$.

Of course, the variational solution by using Gaussians is not optimal, in contrast to the exact solution for the wave function found for the two-nucleon system. A strict solution of the three and four nucleon bound states is possible using the Faddeev–Yakubovsky technique. Within our variational approach, the bound state and scattering properties are reasonably well reproduced, which is sufficient for the present exploratory calculations.

Now we consider in-medium effects at finite density of nuclear matter. For $A = 3, 4$ calculations using a *Faddeev approach* have been performed in [7]. The shifts of binding energy can also be calculated approximately via perturbation theory. In Fig. 6.1, the shifts of the binding energies of the light clusters ($d, t/h$, and α) in symmetric nuclear matter are shown as a function of density for the temperature $T = 10\,\text{MeV}$.

We find that the clusters become modified if they are imbedded in nuclear matter. Like the single nucleon states that become quasiparticles, the contribution of bound states is no longer characterized by only their binding energies which are known, see [22], but have to be calculated on the basis of the nucleon–nucleon interaction. Neglecting the broadening of the bound states due to reactions, a quasiparticle shift can be introduced for the bound states similar to the single-nucleon quasiparticle shift, as has already been discussed in the cluster-mean field approach.

The in-medium Schrödinger equations (6.36) containing quasiparticle shifts and Pauli blocking terms are derived for clusters with mass number A and charge Z [4, 5]. The shift of the bound state energies $E^{\text{qu}}_{A,Z}(P; T, n_i) = E_{A,Z} + \Delta^{\text{SE}}_{A,Z} + \Delta^{\text{Pauli}}_{A,Z} + \Delta^{\text{Coul}}_{A,Z}$, containing in addition to the single-particle self-energy shift $\Delta^{\rm SE}_{A,Z}$ and the Pauli blocking term $\Delta^{\rm Pauli}_{A,Z}$ also the Coulomb shift $\Delta^{\rm Coul}_{A,Z}$, can be calculated within perturbation theory. Besides the quasi-particle shift at zero momentum $\Delta^{\rm SE,0}_{A,Z} = ZE^{\rm qu}_p(0) + (A-Z)E^{\rm qu}_n(0)$, which can be included into the chemical potential, there is the contribution due to the effective mass m^*. Assuming for light clusters $2 \le A \le 4$ a Gaussian wave function with the nucleonic root mean square (r.m.s) radii $\langle r^2\rangle_{A,Z}$, the self-energy shift results as $\Delta^{\rm SE}_{A,Z} = \Delta^{\rm SE,0}_{A,Z} + 3(A-1)\hbar^2 b^2_{A,Z}(m-m^*)/(8m^2)$ after introducing Jacobian coordinates and separating the c.o.m. motion, where $b^2_{A,Z} = 3(A-1)/(A\langle r^2\rangle_{A,Z})$. With Gaussian wave functions, we obtain the following estimate for the Pauli blocking shift

$$\Delta^{\rm Pauli}_{A,Z} = \int \frac{q^2 {\rm d}q e^{-\frac{2Aq^2}{b^2(A-1)}}\left[|E| + \frac{\hbar^2}{2m}\left(\frac{A}{A-1}q^2 + \frac{3(A-2)}{4}b^2\right)\right]}{\left(e^{\left[\frac{\hbar^2}{2m}\left(\frac{P}{A}+q\right)^2 - \hat{\mu}_i\right]/T}+1\right)\left[\frac{A-1}{A}\right]^{3/2}\frac{\sqrt{\pi}}{8}b^3}. \tag{6.66}$$

Similar expressions [20] can be given for the weakly bound clusters with $5 \le A \le 11$. The Coulomb shift $\Delta^{\rm Coul}_{A,Z}$ is calculated in Wigner–Seitz approximation, see also [18]. Within the parameter values considered below, the influence of the Coulomb corrections on the composition is small.

For heavier clusters with mass numbers $A \ge 12$, the self-energy and Pauli-blocking shifts become less important and will be neglected here. The heavier clusters repel the nuclear matter so that the mean-field effects are mostly produced by the other nucleons within the cluster and are contained in the cluster binding energy. For a more detailed consideration see [20].

6.9 Composition of Normal Nuclear Matter

Above the critical temperature $T_{\rm c}$ for the transition to the superfluid state, the approach given here allows to calculate the composition of nuclear matter as well as the thermodynamic properties. We define the mass fractions of the different constituents as

$$X_{A,Z} = A\, n_{A,Z}/n_{\rm B}, \tag{6.67}$$

that is, $X_n = n_n/n_{\rm B}$, $X_p = n_p/n_B$, $X_d = 2\,n_d/n_{\rm B}$, $X_t = 3\,n_t/n_{\rm B}$, $X_h = 3\,n_h/n_{\rm B}$ and $X_\alpha = 4\,n_\alpha/n_{\rm B}$, where $n_{\rm B} = \sum_{A,Z} A\, n_{A,Z}$ is the total baryon density. Furthermore, we introduce the total proton fraction as $Y^{\rm tot}_p = \sum_{A,Z} Z\, n_{A,Z}/n_{\rm B}$.

In thermal equilibrium, within a quantum statistical approach a mass action law can be derived. The densities of the different components are determined by the chemical potentials μ_p and μ_n and temperature T. The densities of the free protons and neutrons as well as of the bound states follow in the nonrelativistic case as (c.f. (6.6))

$$n_{A,Z} = \frac{g_{A,Z}}{2\,\pi^2}\int_0^\infty {\rm d}P\, P^2 \frac{1}{e^{\frac{E_{A,Z}(P;T,\mu_p,\mu_n) - Z\mu_p - (A-Z)\mu_n}{k_{\rm B}T}} - (-1)^A}, \tag{6.68}$$

where for deuterons $g_d = 3$, for tritons $g_t = 2$, for helions $g_h = 2$, and for α particles $g_\alpha = 1$. In the low-density limit where the medium effects can be neglected, the energies $E_{A,Z}(P;T,\mu_p,\mu_n) = m_{A,Z} + P^2/(2m_{A,Z})$ can be used, where $m_{A,Z} = Zm_p + (A-Z)m_n + E^{\rm b}_{A,Z}$ and the binding energies $E^{\rm b}_{A,Z}$ are given in Table 6.1. Note that now the chemical potentials are shifted when including the rest mass of nucleons in the energy.

In the low-density limit, we reproduce the statistical multi-fragmentation model, which is well established in heavy ion collision theory. Here, only the bound state contributions are considered, which means that an ideal mixture of reacting species is considered in thermal equilibrium. The excited states are taken into account, but any interaction between the clusters is omitted. The distribution function is given by (6.6).

The concept of composition anticipates that clusters are well-defined objects in nuclear matter, which becomes questionable with increasing density. If the A-nucleon spectral function which always is a well-defined quantity shows no clear δ-like signatures of quasiparticles, we cannot identify abundances of the corresponding clusters. Considering a nonequilibrium process where warm dense matter expands, in the quasistatic, adiabatic approximation the bound state part of the in-medium A-nucleon spectrum will transform to nuclei, whereas in the sudden approximation a coalescence model can be applied where a projection of the wave function on the free-cluster wave function has to be performed.

Here, we calculate the composition replacing the isolated cluster binding energies by the density dependent ones. We use the estimations for the medium modification of the binding energies given in [21, 23, 24]. Results for the composition of nuclear matter at temperature $T = 10\,\mathrm{MeV}$ with proton fraction $Y_p^{\rm tot} = 0.2$ are shown in Fig. 6.4, for symmetric matter $Y_p^{\rm tot} = 0.5$ in

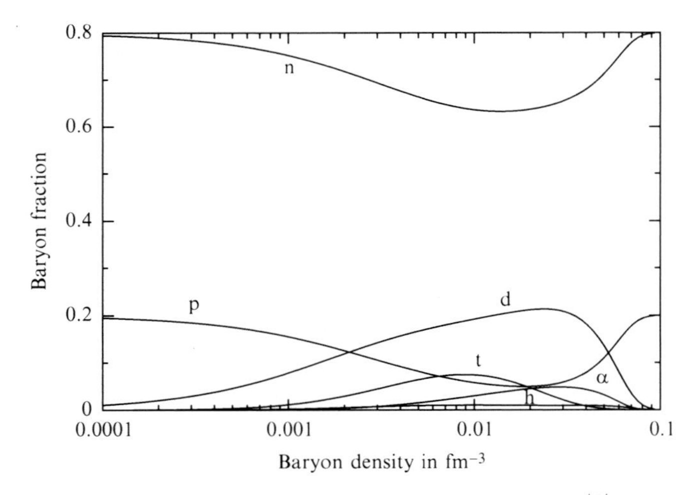

Fig. 6.4. Composition of nuclear matter with proton fraction $Y_p^{\rm tot} = 0.2$ as function of the baryon density, $T = 10\,\mathrm{MeV}$

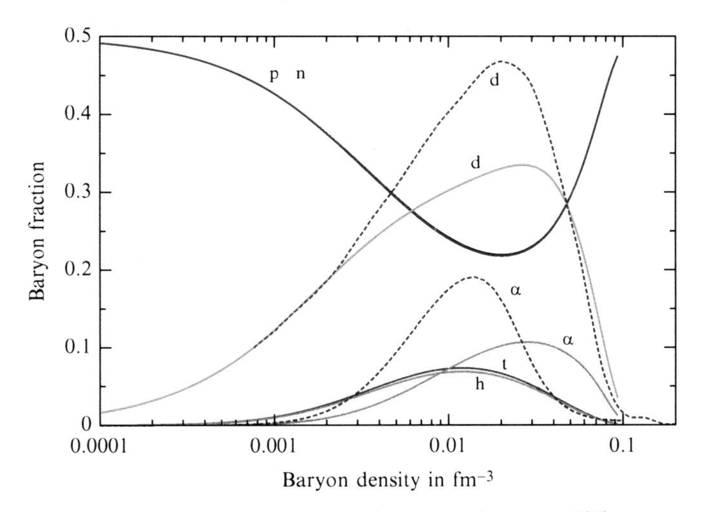

Fig. 6.5. Composition of nuclear matter with proton fraction $Y_p^{\text{tot}} = 0.5$ as function of the baryon density, $T = 10\,\text{MeV}$. *Dashed*, no t, h

Fig. 6.5. The model of an ideal mixture of free nucleons and clusters applies to the low density limit. At higher baryon density, medium effects are relevant. It is shown that in particular α-clusters are formed in symmetric nuclear matter, but they are destroyed at about nuclear matter density. In the case of asymmetric matter, tritons become abundant compared with ^{3}He if $\alpha = 1 - 2Y_p^{\text{tot}} > 0$.

In detail, up to densities of about $0.001\,\text{fm}^{-3}$, density effects can be neglected. This way we describe an ideal mixture in chemical equilibrium. The composition as well as the thermodynamic functions can be calculated immediately by solving the equations given above. Also, the β-equilibrium can be calculated describing the chemical equilibrium with respect to the weak decay $n \rightleftharpoons p + e + \bar{\nu}_\text{e}$, where one usually neglects the chemical potential of the neutrinos. For the electron chemical potential, the relativistic ideal fermion gas model is used. Neglecting the formation of clusters, the corresponding results for the proton fraction as well as the thermodynamical functions are well known from the literature, see [12], where at higher densities a quasiparticle picture is introduced. They are used to describe nuclear matter in β-equilibrium to calculate the structure of neutron stars. There is an additional relation between the chemical potentials of the proton, neutron, and electron due to the charge neutrality condition so that $n_\text{e} = Y_p^{\text{tot}}\, n_\text{B}$. The calculation of nuclear matter in β-equilibrium is improved by taking the formation of light clusters into account. The calculations within the model of an ideal mixture of different components is straightforward, see [21, 23, 24]. Of interest is the influence of the cluster formation on the proton fraction Y_p^{tot} in β-equilibrium. We expect that the formation of clusters will increase the proton fraction.

We conclude that not only the α-particle but also the other light clusters contribute significantly to the composition. This is of relevance for the collapse of the pre-supernovae core because of the influence of nucleonic correlations on the emission and absorption of neutrinos, which determine cooling and heating rates in the evolution of the neutron star [12].

Furthermore, they also contribute to the baryon chemical potential and this way to the modification of the phase instability region with respect to the parameter values temperature, baryon density, and asymmetry. As an example, for symmetric matter the baryon chemical potential as a function of density for $T = 10\,\text{MeV}$ is shown in Fig. 6.6, where the reduction of the instability region is shown when correlations, in particular cluster formation, are taken into account.

Another interesting point is the determination of the symmetry energy. In the low-density limit, the calculation using the cluster–virial expansions [14] show a significant influence of the cluster formation on the symmetry energy. This has recently been confirmed by experiments [25], see Fig. 6.7. A challenging problem is the transition from the low-density limit where the virial expansion is applicable to the saturation density region $n_\text{B} \approx n_0 = 0.17\,\text{fm}^{-3}$ where the mean-field approaches can be used. The present approach is able to solve this problem.

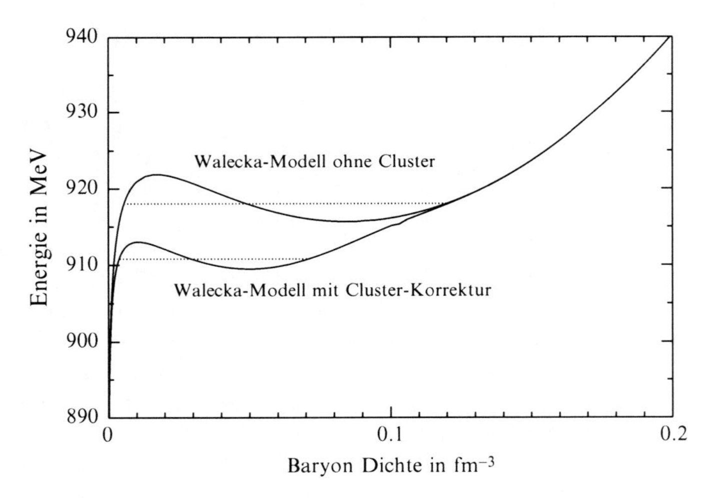

Fig. 6.6. Baryon chemical potential μ (axis of ordinate in MeV) as function of the baryon density n_B (abscissa in fm^{-3}) for symmetric nuclear matter, $T = 10\,\text{MeV}$. Without (*upper case*) and including (*lower case*) the formation of light clusters. The region of phase instability is obtained from a Maxwell construction

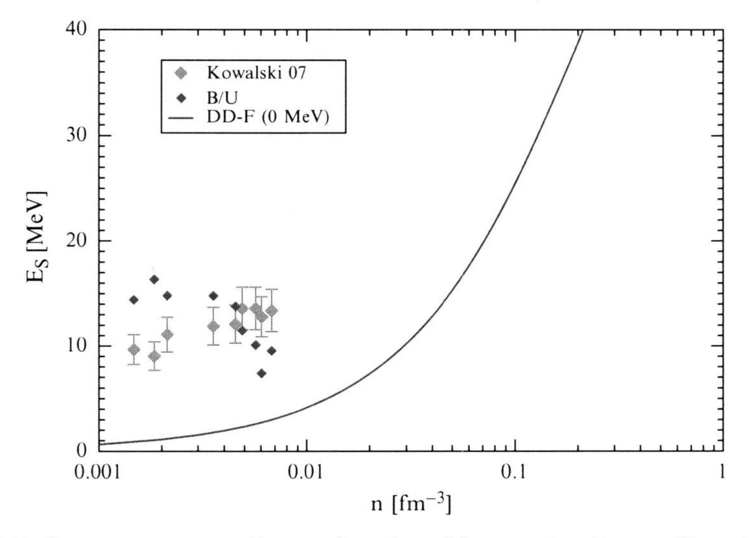

Fig. 6.7. Symmetry energy E_s as a function of baryon density n_B. Experimental values (Kowalski 07 [25]) are compared with the cluster–virial expansion (B/U [14]) and the relativistic mean field approach (DD-F (0 MeV) [2])

6.10 Comparison with the Concept of Excluded Volume

To get the correct physical behavior, medium modifications for the clusters have to be taken into account at high densities. A simple approach is the concept of an excluded volume as used in [12, 18]. Other effects such as the modification of quasi-particles forming a bound state are not considered. For details we refer to [12]. We have also shown in Fig. 6.5 the result for the composition if only α-particle formation is taken into account, using the concept of the excluded volume. The abundance of the α-particle increases up to baryon densities of about a tenth of nuclear matter density and is rapidly decreasing with higher densities. In contrast, the quantum statistical approach shows a more weak decrease of the correlated density with the baryon density. In particular, two-particle correlations are present up to nuclear matter density. Discussing the difference in both approaches, we first note that the concept of a hard core that leads to the excluded volume overestimates the Pauli blocking, which makes the interaction potential more softer. Further, in addition to the medium modification due to the Pauli blocking, the effect of the quasiparticle self-energy shift has to be taken into account.

6.11 Two-Particle Condensates at Low Temperatures

The evaluation of the Beth–Uhlenbeck formula including two-particle correlations has been carried out in [6] based on a separable nucleon–nucleon potential. The result [8] for the composition of nuclear matter as a function of density and temperature is shown in Fig. 6.3. Two aspects of this study of two-particle condensation deserve special attention.

1. The contribution of the correlated density, which derives both from deuterons as bound states in the isospin-singlet channel and from scattering states, is found to increase with decreasing temperature, in accordance with the law of mass action. This law also predicts the increase of correlated density with increasing nucleon density (as also seen in Fig. 6.3 for the low-density limit).

However, under increasing density, the binding energy of the bound state (deuteron) decreases due to Pauli blocking (*Mott effect*). At the Mott density, introduced above, the bound states with vanishing c.o.m. momentum are dissolved in the continuum of scattering states. Bound states with higher c.o.m. momentum merge with the continuum at higher densities. According to Levinson's theorem, if a bound state merges with the continuum, the scattering phase shift in the corresponding channel exhibits a jump by π, such that no discontinuity appears in the EoS. Accordingly, the contribution of the correlated density will remain finite at the Mott density, but will be strongly reduced at somewhat higher densities.

Thus, one salient result is the disappearance of bound states and correlated density already below the saturation density of nuclear matter. The underlying cause of the Mott effect is Pauli blocking, which prohibits the formation of bound states if the phase space is already occupied by the medium (Fermi sphere), and hence no longer available for the formation of the wave function of the bound state (momentum space). This effect holds also for higher-A bound states such as the triton, helion, and α particle, which disappear at corresponding densities (see Fig. 6.1).

2. The Bose pole in the correlated density signals the onset of a quantum condensate. As it is well known, for the bound-state (deuteron channel) contribution, the T-matrix approach breaks down when the pole corresponding to the bound-state energy coincides with twice the chemical potential. In the following, we restrict us to symmetric matter, $\mu_p = \mu_n = \mu$. The Thouless condition, embodied in

$$\begin{aligned}\psi_{2,\nu,P}(12) &= \sum_{1'2'} \frac{1 - \tilde{f}_1(1) - \tilde{f}_1(2)}{2\mu - E^{\mathrm{qu}}(1) - E^{\mathrm{qu}}(2)} \psi_{2,\nu,P}(1'2') \\ &= \sum_{1'2'} K_2(12, 1'2', 2\mu)\psi_{2,\nu,P}(1'2')\end{aligned} \tag{6.69}$$

describes the onset of a quantum condensate. The same condition also holds for the contribution of scattering states. Consequently, the transition temperature for the onset of a quantum condensate appears as a smooth function of density, as shown in Fig. 6.8.

Below the transition temperature, the T-matrix approach is no longer applicable. However, a mean-field approach becomes possible in this regime after performing a Bogoliubov transformation. Even so, the proper inclusion

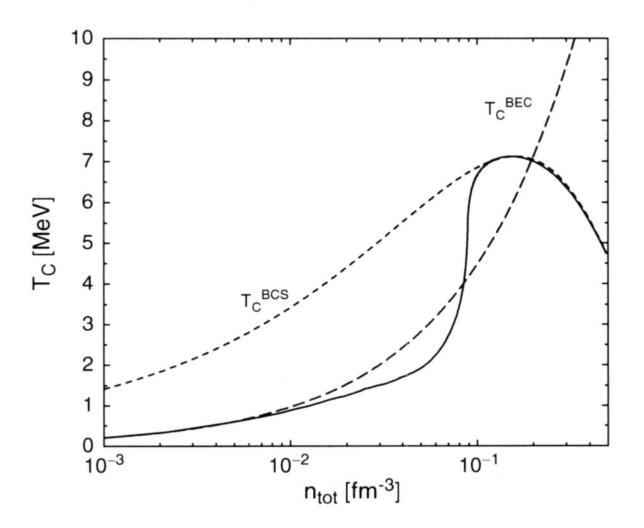

Fig. 6.8. Onset T_c of quantum condensate in the two-particle channel in symmetric nuclear matter as a function of the total baryon density $n_{\mathrm{tot}} = n_{\mathrm{B}}$. Crossover from BEC to BCS occurs as the density increases

of correlations below the critical temperature remains a challenging problem. To date, only the first steps have been taken [26] toward solving this problem for general quantum many-particle systems.

Another important aspect of the problem of two-particle condensation, indeed one of great current interest, is the interpretation of the critical temperature. At low densities, where the two-body bound states (deuterons) are well-defined composite particles, the mass–action law implies that the deuterons will dominate the composition in the low-temperature region. In this region, the critical temperature for the transition to the quantum condensate coincides with the BEC of deuterons as known for ideal Bose systems. At high densities, where bound states are absent, the transition temperature coincides with the solution of the Gor'kov equation describing the formation of Cooper pairs. Thus, BEC and BCS scenarios characterize the low- and high-density regimes, respectively. We observe a smooth *crossover transition* from BEC to BCS behavior [8] that is general issue in fermion systems, for example, in [27], and is currently the subject of intense experimental study in cold atomic gases.

Going beyond the mean-field approximation, the first remarkable feature [28] emerging at the two-particle level is the formation of a *pseudogap* in the density of states (DoS) above the critical temperature T_c. Compared with the orthodox BCS solution, for which a gap opens in the DoS below T_c, a quite different situation is present in strongly correlated Fermi systems. The full treatment of the (two-body) T-matrix leads to a reduction of the DoS near the

Fermi energy already *above* T_c, within an energy interval of the same order as the BCS gap at zero temperature. This behavior may be traced to fluctuations above T_c that presage the transition to the superfluid state. Similar precursor behavior is known to occur in other systems of strongly correlated fermions. In the Hubbard model, for example, the formation of local magnetic moments already begins above the critical temperature at which long-range order of the moments becomes manifest. The pseudogap phenomenon is of course a widely discussed aspect of compounds exhibiting high T_c superconductivity [29].

In the context of nuclear matter, the occurrence of a pseudogap phase was first considered by Schnell et al. [28] in the quasiparticle approximation, as noted earlier. They showed that this effect is partially washed out if a self-consistent approximation for the spectral function is implemented, but a full description should take vertex corrections into account. A similar assessment applies to a recent self-consistent solution of the Gor'kov equation in terms of the spectral function [30], which shows a reduction of the transition temperature for quantum condensation. However, vertex corrections should also be included in this case and may partially compensate the self-energy effects.

6.12 Four-Particle Condensates and Quartetting in Nuclear Matter

In general, it is necessary to take account of *all bosonic clusters* to gain a complete picture of the onset of superfluidity. The picture developed in the preceding section includes only the effects of two-particle correlations leading to two-body deuteron clusters. However, as is well known, the deuteron is weakly bound compared to other nuclei. Higher-A clusters can arise that are more stable. In this section, we consider the formation of α particles, which are of special importance because of their large binding energy per nucleon (7 MeV). We will not include tritons or helions, which are fermions and not so tightly bound. Moreover, we will not consider nuclei in the iron region, which have even larger binding energy per nucleon than the α and thus comprise the dominant component at low temperatures and densities. The latter are complex structures of many particles and are strongly affected by the medium as the density increases, so that they are assumed to be not of relevance in the temperature and density region considered here.

The in-medium wave equation for the four-nucleon problem has been solved using the Faddeev–Yakubovski technique, with the inclusion of Pauli blocking. The binding energy of an α-like cluster with zero c.o.m. momentum vanishes at around $n_0/10$, where $n_0 \simeq 0.17\,\text{nucleons}\,\text{fm}^{-3}$ denotes the saturation density of isospin-symmetric nuclear matter. Thus, the four-body bound states make no significant contribution to the composition of the system above this density. Given the medium-modified bound-state energy $E^{\text{qu}}_{4,\alpha}(P)$, the bound-state contribution to the EoS is

$$n_\alpha(\beta,\mu) = \sum_P \left[e^{\beta(E^{\mathrm{qu}}_{4,\alpha}(P) - 2\mu_p - 2\mu_n)} - 1 \right]^{-1} . \tag{6.70}$$

We will not include the contribution of the excited states or that of scattering states. Because of the large specific binding energy of the α particle, low-density nuclear matter is predominantly composed of α particles. This observation underlies the concept of α matter and its relevance to diverse nuclear phenomena.

Symmetric nuclear matter is characterized by equality of the proton and neutron chemical potentials, that is, $\mu_p = \mu_n = \mu$. The four-particle correlations are embodied in the *four-particle Green function*, which in the ladder approximation is given by

$$\begin{aligned} G_4(1234, 1'2'3'4', \Omega_4) &= \frac{\tilde{f}_1(1)\tilde{f}_1(2)\tilde{f}_1(3)\tilde{f}_1(4)}{\tilde{f}_4(E_4(1234))} \frac{\delta_{11'}\delta_{22'}\delta_{33'}\delta_{44'}}{\Omega_4 - E_4(1234)} \\ &+ \sum_{1''2''3''4''} K_4(1234, 1''2''3''4'', \Omega_4) G_4(1''2''3''4'', 1'2'3'4', \Omega_4) , \end{aligned} \tag{6.71}$$

where $E_4(1234) = E^{\mathrm{qu}}(1) + E^{\mathrm{qu}}(2) + E^{\mathrm{qu}}(3) + E^{\mathrm{qu}}(4)$ and $\tilde{f}_4(E)$ is the Bose distribution function with the effective chemical potential $\tilde{\mu}$. The interaction kernel K is obtained using the technique of the Matsubara Green functions, which yields

$$\begin{aligned} K_4(1234, 1'2'3'4', \Omega_4) &= V(12, 1'2')\delta_{33'}\delta_{44'} \frac{\tilde{f}(1)\tilde{f}(2)}{\tilde{f}_2(E_2(12))} \frac{1}{\Omega_4 - E_4(1234)} \\ &\times \left[1 + \tilde{f}_4(1'2'34) \frac{\Omega_4 - E_4(1'2'34)}{E_2(12) - E_2(1'2')} \left(e^{(E_2(1'2') - 2\mu)/T} - e^{(E_2(12) - 2\mu)/T} \right) \right] \\ &+ \cdots + \cdots + \cdots + \cdots + \cdots , \end{aligned} \tag{6.72}$$

where the terms obtained by relabeling are not shown explicitly. In the zero-density limit $n_{\mathrm{B}} \to 0$, where the Fermi distribution function is small compared with unity, this becomes simply

$$\lim_{n_B \to 0} K_4(1234, 1'2'3'4', \Omega_4) = \frac{\delta_{11'}\delta_{22'}\delta_{33'}\delta_{44'}}{\Omega_4 - E_4(1234)} \sum_{i<j} V(ij, i'j')\delta_{kk'} \tag{6.73}$$

in terms of the bare interaction term of the four-body system.

The poles of the analytic continuation of $G_4(1234, 1'2'3'4', z)$ into the complex z plane are of special interest. Near the pole at $E^{\mathrm{qu}}_{4,\nu}(P)$, the Green function can be factorized as

$$G_4(1234, 1'2'3'4', z) \approx \psi_{4,\nu,P}(1234)\psi^*_{4,\nu,P}(1'2'3'4')/(z - E^{\mathrm{qu}}_{4,\nu}(P)) . \tag{6.74}$$

In this expression, $E^{\mathrm{qu}}_{4,\nu}(P)$ and $\psi_{4,\nu,P}(1234)$ are the eigenvalues and the eigenstates of the four-particle system, which follow from the solution of the four-particle Schrödinger-like wave equation

$$\psi_{4,\nu,P}(1234) = \sum_{1'2'3'4'} K_4(1234, 1'2'3'4', E^{\mathrm{qu}}_{4,\nu}(P))\psi_{4,\nu,P}(1'2'3'4') \,. \qquad (6.75)$$

Importantly, bound states can exist, and we denote the lowest bound state by $E_{4,0}$, which in the nuclear context is the α particle. Because of the influence of the medium as reflected in the self-energy and the phase-space occupation factors, the bound-state energy depends on the temperature T and chemical potential μ.

The four-particle correlations as contained in the four-particle Green function G_4 serve to determine the thermodynamic properties of the system, including the EoS. For example, the *four-particle density* is given by

$$\langle a_1^\dagger a_2^\dagger a_3^\dagger a_4^\dagger a_{4'} a_{3'} a_{2'} a_{1'} \rangle = \int \frac{\mathrm{d}\omega}{\pi} f_4(\omega) \mathrm{Im} G_4(1234, 1'2'3'4', \omega - i0) \,. \qquad (6.76)$$

Obviously, this density diverges when, at a given temperature, the chemical potential attains the value $4\mu = E^{\mathrm{qu}}_{4,\nu}(0)$. At that point, the delta function produced by the pole coincides with the singularity of the Bose distribution function. This singularity is directly related to the onset of superfluidity [9]; for example, at low densities it will lead to Bose condensation of α particles. In general, the condition for the onset of superfluidity due to four-particle correlations follows from the equation

$$\psi_{4,\alpha,0}(1234) = \sum_{1'2'3'4'} K_4(1234, 1'2'3'4', 4\mu)\psi_{4,\alpha,0}(1'2'3'4') \,, \qquad (6.77)$$

which determines the critical temperature $T_{\mathrm{c}}(\mu)$. It should be noted that analogous arguments are used to determine the onset of pairing by considering the behavior of the two-particle propagator [31].

The results in solving (6.77) are presented in Figs. 6.9 and 6.10. An important consequence is that at the lowest temperatures, BEC occurs for α particles rather than deuterons. As the density increases within the low-temperature regime, the chemical potential μ first reaches $-7\,\mathrm{MeV}$, where the α's Bose-condense. By contrast, Bose condensation of deuterons would not occur until μ rises to $-1.1\,\mathrm{MeV}$.

The calculation reveals that in the low-density region, the critical density tracks that for BEC of ideal α particles; hence the Bose condensation of deuterons as considered in the previous section becomes irrelevant. As expected, with increasing density the transition temperature deviates from that of the ideal Bose gas of α's due to medium corrections.

Moreover, the *quartetting* transition temperature is sharply reduced as the rising density approaches the critical Mott value at which the four-body bound states disappear. At that point, pair formation in the isospin-singlet deuteron-like channel comes into play, and a deuteron condensate will exist below the critical temperature for BCS pairing up to densities above the nuclear-matter saturation density $n_0 \approx 0.17\,\mathrm{fm}^{-3}$, as described in the previous section. The critical density at which the α condensate disappears is

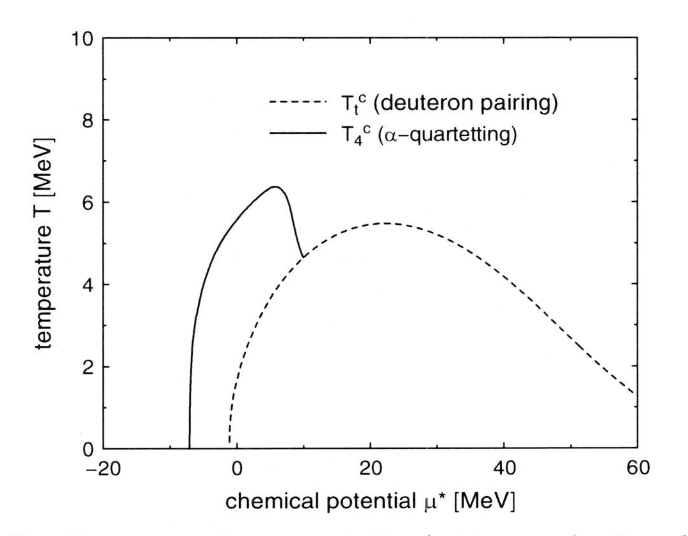

Fig. 6.9. Transition temperature to quartetting/pairing as a function of chemical potential μ^* (without single particle mean-field shift) in symmetric nuclear matter

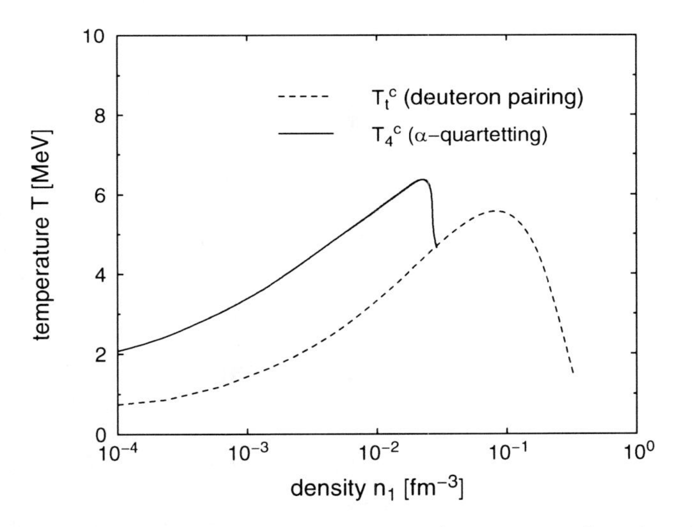

Fig. 6.10. Transition temperature to quartetting/pairing as a function of the free nucleon density n_1 in symmetric nuclear matter

estimated to be $n_0/3$. However, the variational approach of [9] on which this estimate is based represents only a first attempt of describing the transition from quartetting to pairing. The detailed nature of this fascinating transition remains to be clarified.

The BEC for ideal quantum gases is a well-known phenomenon. The occupation of single-particle states is given by the Bose distribution function.

Below a critical temperature T_c, to obey normalization, the state of lowest energy is macroscopically occupied. This macroscopically enhanced coherent occupation of the lowest quantum state is denoted as quantum condensate. As well known, the fraction of bosons found in the condensate results for the ideal Bose gas as $n_{\rm cond}/n = 1 - (T/T_c)^{3/2}$.

However, this simple picture is no longer valid if interaction is taken into account. Here, we want to concentrate on interaction effects at zero temperature. In general, the condensate fraction is given by the properties of the density matrix, which contains a part which factorizes. According to Penrose and Onsager [32], the quantum condensate in a homogeneous interacting boson system at zero temperature is given by the off-diagonal long-range order in the density matrix. The nondiagonal density matrix in coordinate representation can be factorized so that in the limit $|\boldsymbol{r} - \boldsymbol{r}'| \to \infty$ follows

$$\lim_{|\boldsymbol{r}-\boldsymbol{r}'|\to\infty} \rho(\boldsymbol{r}, \boldsymbol{r}') = \psi_0^*(\boldsymbol{r})\psi_0(\boldsymbol{r}') + \gamma(\boldsymbol{r} - \boldsymbol{r}'). \tag{6.78}$$

The last contribution $\gamma(r)$ disappears at large distances, whereas the first contribution determines the condensate fraction in infinite matter as

$$n_0 = \frac{\langle \Psi | a_0^\dagger a_0 | \Psi \rangle}{\langle \Psi | \Psi \rangle}, \tag{6.79}$$

with $a_0^\dagger a_0$ being the occupation number of the condensate state $\psi_0(\boldsymbol{r})$. Exploratory calculation of the condensate fraction of α matter will be given in the following section. In contrast to [9] where the transition temperature T_c for quartetting was considered, we consider here the zero temperature case and analyze the ground state wave function. It will be shown that due to the interaction, the condensate fraction is suppressed with increasing density.

6.13 Suppression of Condensate Fraction in α Matter at Zero Temperature

The theory of Penrose and Onsager [32] was first applied to a system with hard core repulsion. Depending on the filling factor, the suppression of the condensate was calculated. In particular, for liquid ^{4}He with a filling factor of 28% at normal conditions, the condensate fraction is reduced to $\approx$8%, in good agreement with experimental observations. To give an estimation for α matter, with an "excluded volume" of about 20 fm^3 [18], such a filling factor of 28% would arise at $\approx n_0/3$ so that a substantial reduction of the condensate fraction already below saturation densities is expected for α matter.

Within a more systematic approach, we follow the work of Clark et al. [33]. We calculate the reduction of the condensate fraction as function of the baryon density within perturbation theory. A uniform Bose gas of α particles, interacting via the potential $V_\alpha(r)$, is considered, disregarding any change of the internal structure of the α particles at increasing density. In particular, the dissolution of the α particle as a four-nucleon bound state because of the Pauli blocking is not taken into account.

The simplest form of a trial wave function incorporating the strong spatial correlations implied by the interaction potential is the familiar Jastrow choice, $\psi(\boldsymbol{r}_1,\ldots,\boldsymbol{r}_A)=\prod_{i<j} f(|\boldsymbol{r}_i-\boldsymbol{r}_j|)$. Within our exploratory calculation, we consider the lowest approximation with respect to the density to show the tendency of condensate suppression due to the interaction. Normalization gives for the variational function the constraint

$$4\pi n_\alpha \int_0^\infty [f^2(r)-1]r^2\mathrm{d}r = -1, \tag{6.80}$$

$n_\alpha = n_\mathrm{B}/4$ being the density of α particles.

In the low density limit, the binding energy per α-particle is given by

$$E[f] = 2\pi n_\alpha \int_0^\infty \left\{ \frac{\hbar^2}{4m}\left(\frac{\partial f(r)}{\partial r}\right)^2 + V_\alpha(r) f^2(r) \right\} r^2 \mathrm{d}r, \tag{6.81}$$

m being the nucleon mass. The condensate fraction is calculated according to

$$n_0 = \exp\left\{ -4\pi n_\alpha \int_0^\infty [f(r)-1]^2 r^2 \mathrm{d}r \right\}. \tag{6.82}$$

Note that these approximations [33] only hold in the low-density limit. At higher densities, the pair correlation function has to be evaluated. A more advanced approach based on a HNC calculation has been given by Clark, Ristig, and others, see [33, 34].

For the evaluation of the condensate fraction (6.82), we use the Ali–Bodmer α–α interaction potential [35]

$$V_\alpha(r) = 457\,\mathrm{e}^{-(0.7r/\mathrm{fm})^2}\mathrm{MeV} - 130\,\mathrm{e}^{-(0.475r/\mathrm{fm})^2}\mathrm{MeV}. \tag{6.83}$$

According to Johnson and Clark [33], we choose the variational function as

$$f(r) = (1-\mathrm{e}^{-ar})(1+b\,\mathrm{e}^{-ar}+c\,\mathrm{e}^{-2ar}). \tag{6.84}$$

After determining the parameters a, b, c from the minimum of energy [36], the condensate fraction can be evaluated, see Fig. 6.11.

In Fig. 6.11, the full line represents the result for the condensate fraction as function of the baryonic density according to the perturbative treatment. In the zero density limit this fraction is expected to go to 1. Calculations performed by Johnson and Clark [33] using a HNC calculation for the pair distribution function are given by crosses, showing a stronger suppression of the condensate fraction near the saturation density.

As found from the calculation of the critical temperature for the formation of a quartetting condensate [9], we expect that the condensate fraction will

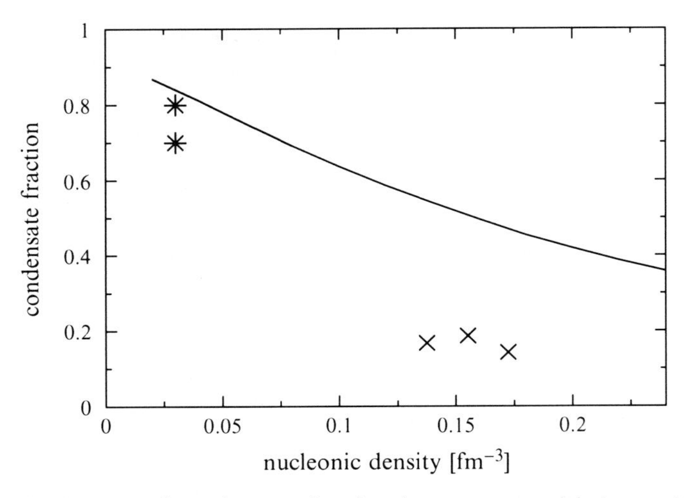

Fig. 6.11. Reduction of condensate fraction in α matter with increasing baryon density n_B. *Full line*, perturbation approach; *crosses*, HNC calculations by Johnson and Clark [33]; *stars*, Hoyle state (see Sect. 6.14)

disappear near the saturation density. For this, we have not only to take into account the HNC type improvement of the pair distribution function, but also the Pauli blocking effects that modify the internal structure of the α particle so that the use of the Ali–Bodmer interaction potential is no longer justified. Recently, improved versions of the α–α interaction have been proposed [37]. Three-α forces have been considered to give a better estimation for the critical point of α matter, which should be positioned below saturation density. Thus, the repulsive part of the α–α interaction (which also is a consequence of the Pauli blocking with respect to the internal nucleonic structure) is only a part of the suppression of the condensate, which is described here.

Another effect is the medium modification of the internal structure of the α particle as well as of the interaction that can be elaborated within a cluster-mean field approximation [4]. The dissolution of α-like bound states due to Pauli blocking has been evaluated for an uncorrelated medium solving the Faddeev–Yakubowsky equation [7]. It has been shown [9] that the four-particle correlations in the condensate disappear due to Pauli blocking at around $n_0/3$ within a variational approach, approximating the four-nucleon wave function by the solution of the two-particle problem and describing the relative c.o.m. motion by a Gaussian wave function. Therefore, a medium-dependent α–α interaction of the Ali–Bodmer type may be expected to account for the features of this effect in an exploratory way. In principle, an ab initio calculation based on interacting nucleons should be performed, with Green functions, variational, or AMD techniques.

6.14 Enhancement of Cluster c.o.m. *S* Orbital Occupation in $4n$ Nuclei

An important question is whether such properties of infinite nuclear matter are of relevance for finite nuclei. As well known, for example, pairing obtained in nuclear matter within the BCS approach is also clearly seen in finite nuclei. Nuclei with densities near the saturation density are well described by the quasiparticle picture, which leads to the shell model for finite nuclei. At low densities, a fully developed α cluster structure similar to α matter is expected. Cluster structures in finite nuclei have been well established. A density functional approach is able to include correlations and to bridge between infinite matter and finite nuclei.

An interesting aspect of finite nuclei is the enhancement of the occupation of single α-particle states similar to BEC in α-particle matter or condensation of bosonic atoms in traps. Recently, gas-like states have been investigated in self-conjugate $4n$ nuclei [38], and a special ansatz for the wave function (THSR ansatz), which is similar to the condensate state in infinite matter, has been shown to be appropriate in describing low-density isomers. In particular, ^{8}Be and the Hoyle state of ^{12}C are well described with this THSR wave function. Investigations of states near the four α threshold in ^{16}O are in progress [39,40].

Signatures akin to BEC calculated for infinite nuclear matter should arise already in finite nuclei. Low-density states of self-conjugate $4n$ nuclei clearly show an α cluster structure, in particular for $n = 2$ and $n = 3$ (Hoyle state). The counterpart of a condensate in infinite α matter, where the occupation of the ground state is enhanced and becomes of the same order as the total particle number, will be the enhancement of the occupation number of a single-α orbital of the α-clusters in a low density state of the nucleus.

The α clustering nature of the nucleus ^{12}C has been studied by many authors using various approaches [41]. Among these studies, solving the fully microscopic three-body problem of α clusters gives us the most important and reliable theoretical information of α clustering in ^{12}C within the assumption that no α cluster is distorted or broken, except for the change of the size parameter of the α cluster's internal wave function. First solutions of the microscopic 3α problem where the antisymmetrization of nucleons is exactly treated have been given by Uegaki et al. [42] and by Kamimura et al. [43]. In those works, the ^{12}C levels are described by the wave function of the form $\mathcal{A}\{\chi(\boldsymbol{s},\boldsymbol{t})\phi_\alpha^3\}$ with $\mathcal{A}$ standing for the antisymmetrizer, $\phi_\alpha^3 \equiv \phi(\alpha_1)\phi(\alpha_2)\phi(\alpha_3)$ for the product of the internal wave functions of three α clusters, and $\boldsymbol{s}$ and $\boldsymbol{t}$ for the Jacobi coordinates of the center-of-mass motion of three α clusters. Here $\phi(\alpha_i)$ $(i = 1, 2, 3)$ is the internal wave function of the α-cluster α_i having the form $\phi(\alpha_i) \propto \exp[-(1/8b^2)\sum_{m>n}^{4}(\boldsymbol{r}_{i,m} - \boldsymbol{r}_{i,n})^2]$. The wave function $\chi(\boldsymbol{s},\boldsymbol{t})$ of the relative motion of three α clusters is obtained by solving the energy eigenvalue problem of the full three-body equation of motion; $\langle\phi_\alpha^3|(H - E)|\mathcal{A}\{\chi(\boldsymbol{s},\boldsymbol{t})\phi_\alpha^3\}\rangle = 0$, where H is the microscopic Hamiltonian consisting of the kinetic energy, effective two-nucleon potential, and the Coulomb potential between protons.

Both calculations reproduce reasonably well the observed binding energy and r.m.s. radius of the ground 0_1^+ state, which is the state with normal density, while they both predict a very large r.m.s. radius for the second 0_2^+ state, which is larger than the r.m.s. radius of the ground 0_1^+ state by about 1 fm, that is, by over 30%. The second 0^+ state of ^{12}C is well known as the key state for the synthesis of ^{12}C in stars (Hoyle state) and also as one of the typical mysterious 0^+ states in light nuclei, which are very difficult to understand from the point of view of the shell model [44].

Alternatively, the 0_2^+ state with dilute density can be described by a gas-like structure of 3α-particles, which interact weakly with each other, predominantly in relative S waves. The S-wave dominance in the 0_2^+ state structure had been already suggested by Horiuchi on the basis of the 3α OCM (orthogonality condition model) calculation [45]. It should be mentioned that not only the binding energy, but also other properties of the 0_2^+ state such as electron scattering form factors are well described within the calculations given in [42, 43, 45].

Recently, based on the investigations of the possibility of α-particle condensation in low-density nuclear matter [9], it was conjectured that, near the $n\alpha$ threshold in self-conjugate $4n$ nuclei, there exist excited states of dilute density, which are composed of a weekly interacting gas of self-bound α particles and which can be considered as an $n\alpha$ condensed state [38]. The structure of ^{12}C and ^{16}O was examined using a new α-cluster wave function of the α-cluster condensate type. The new α-cluster wave function, which will be denoted as THSR wave function, actually succeeded to place an excited state of dilute density (about one third of saturation density) in ^{12}C and ^{16}O at energy in the vicinity of the three, respectively, 4 α breakup threshold, without using any adjustable parameter.

The THSR wave function of the α-cluster condensate type used in [38] represents a condensation of α-clusters in a spherically symmetric state. This is clearly seen by the following expression

$$|\Psi\rangle = \mathcal{P}(C_\alpha^\dagger)^n|\text{vac}\rangle, \tag{6.85}$$

with

$$\langle 1234|C_\alpha^\dagger|\text{vac}\rangle = \Phi(\boldsymbol{P})\delta_{\boldsymbol{P},\boldsymbol{p}_1+\boldsymbol{p}_2+\boldsymbol{p}_3+\boldsymbol{p}_4}\phi_\alpha(1234)a_1^\dagger a_2^\dagger a_3^\dagger a_4^\dagger, \tag{6.86}$$

$\Phi(\boldsymbol{P})$ describing the c.o.m. motion of the α cluster, and ϕ the internal wave function of the four-nucleon cluster. The operator $\mathcal{P}$ is projecting out the total c.o.m. motion of the $4n$ nucleus. In the limit of infinite nuclear matter, the Φ orbitals are plane waves, and the projection operator $\mathcal{P}$ can be neglected. In the case considered here, the use of Gaussians allows the explicit separation of the c.o.m. motion of the four-nucleon cluster as well as of the whole $4n$ nucleus. It should also be noted that (6.85) contains two limits exactly: the one of a pure Slater determinant relevant at higher densities and the one of a 100% ideal α-particle condensate in the dilute limit [38].

This wave function (6.85) has been extended so that it can describe the α-cluster condensate with spatial deformation [46]. This new wave function, applied to ^{8}Be, succeeded to reproduce not only the binding energy of the ground state but also the energy of the excited 2^+ state.

It was shown that the 0_2^+ wave function of ^{12}C, which was obtained long time ago by solving the full three-body problem of the microscopic 3α cluster model, is almost completely equivalent to the wave function of the 3α THSR state. The rms radius for this THSR state was calculated as $R(0_2^+)_{\text{THSR}} =$ 4.3 fm, which fits well with the experimental data for the form factor of the Hoyle state, see [47]. It confirms the assumption of low density as a prerequisite for the formation of an α-cluster structure for which the Bose-like enhancement of the occupation of the S orbit is possible.

Recently, a fermionic AMD calculation based on nucleons with effective interactions has been performed [47], which supports the applicability of the THSR state to describe the Hoyle state. It is found that the form factor calculated for the 0_2^+ state of ^{12}C coincides with the form factor obtained from the THSR wave function. In particular, the low density of nucleons, the formation of four-nucleon clusters, and the dominant contribution of the gas-like distribution has been confirmed.

A very interesting analysis of the applicability of the THSR wave function can be performed by comparing with stochastic variational calculations [48] and OCM calculations [49]. The α density matrix $\rho(\boldsymbol{r}, \boldsymbol{r}')$ defined by integrating out of the total density matrix all intrinsic α-particle coordinates is diagonalized to study the single-α orbits and occupation probabilities in ^{12}C states. Figure 6.12 shows the occupation probabilities of the S orbits as a function of the nuclear density corresponding to the rms radius, including the

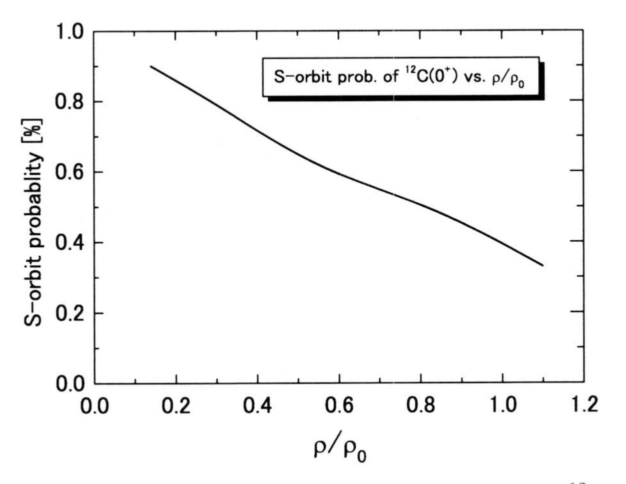

Fig. 6.12. Occupation of the S-wave α orbital ($1.0 = 100\%$) of ^{12}C as a function of the nuclear density $n_{\text{B}}/n_0 \approx \rho/\rho_0$, where the ratio ρ/ρ_0 is determined by the rms radii of the excited nucleus and the ground state nucleus [49]

ground and Hoyle state of ^{12}C, obtained by diagonalizing the density matrix $\rho(\boldsymbol{r}, \boldsymbol{r}')$. In the Hoyle state, the $0S$ α-particle orbit with zero node is occupied to more than 70% by the three α-particles (see also [48]). This huge percentage means that an almost ideal α-particle condensate is realized in the Hoyle state. One should remember that superfluid ^{4}He has only 8% of the particles in the condensate, which represents a macroscopic amount of particles nonetheless. The S-wave occupancy of the Hoyle state results at least by a factor of 10 larger than the occupancy of any other state. On the other hand, in the ground state of ^{12}C, the α-particle occupations are equally shared among $S1$, $D1$, and $G1$ orbits, where they have two, one, and zero nodes, respectively, reflecting the SU(3)$(\lambda\mu)$ = (04) character of the ground state [49].

An interesting item is whether there exist other nuclei showing the Bose condensate-like enhancement of the S-orbit occupation number. Then, the suppression of the condensate with increasing density is also of relevance for those nuclei. In analogy to the aforementioned OCM calculation for ^{12}C [49], recently a quite complete OCM calculation has been performed also for ^{16}O [40]. It was possible to reproduce the full spectrum of 0^+ states with 0_2^+ at 6.4 MeV, 0_3^+ at 9.4 MeV, 0_4^+ at 12.6 MeV, 0_5^+ at 14.1 MeV, and 0_6^+ at 16.5 MeV. Also the rms radii are obtained. The largest values are found as $R(0_6^+)_{\rm OCM} = 5.4\,\rm fm$, followed by $R(0_4^+)_{\rm OCM} = 3.9\,\rm fm$. The analysis of the diagonalization of the α-particle density matrix $\rho(\boldsymbol{r}, \boldsymbol{r}')$ (as was done in [49]) showed that the newly discovered 0^+ state at 13.6 MeV [50], as well as the well known 0^+ state at 14.01 MeV, corresponding to the states at 12.6 MeV and 14.1 MeV, respectively, have, contrary to what we assumed previously [51], very little condensate occupancy of the $0S$-orbit (about 20%). On the other hand, the sixth 0^+ state for which the energy 16.5 MeV has been calculated (to be identified with the experimental state at 15.1 MeV) has 61% of the α particles being in the $0S$-orbit.

These results confirm the statement that the α-particle condensate in nuclear matter is suppressed with increasing density and, consequently, a well developed condensate state in nuclei can be expected only at very low densities. For ^{16}O, the relative densities ρ/ρ_0 are estimated as $(R(0_4^+)_{\rm OCM}/R(0_1^+)_{\rm exp})^3 = 0.34$ and $(R(0_6^+)_{\rm OCM}/R(0_1^+)_{\rm exp})^3 = 0.13$. Therefore, we expect a significant enhancement of the S orbit occupation number only for the 0_6^+ state, in full agreement with the OCM calculation cited above. The very large radius of that state is again a clear indication of an α-particle gas (Hoyle)-like state, and the THSR wave function is expected to describe this state in a sufficient approximation. Work in determining the complete spectrum of THSR states in ^{16}O showing the relevance of a Bose-condensate like state is in progress [39].

6.15 Conclusions

In certain regions of the density–temperature plane, a significant fraction of nuclear matter is bound into clusters. For instance, at low densities deuterons can be formed as a two-nucleon bound state. The mass fraction of nucleons bound into deuterons is given by a mass action law. Macroscopic properties such as the EoS are influenced by the occurrence of bound states. Within a consistent description, also scattering states have to be included, which can be done within a quantum statistical approach such as the Beth–Uhlenbeck formula.

With increasing density, the single-particle properties as well as the two-particle properties are modified by medium effects. The quantum statistical approach allows to calculate self-energy and Pauli blocking. The single-nucleon as well as the two-nucleon spectral function allow for a comprehensive and exhaustive description of all correlation effects. Neglecting the broadening of the states, a quasiparticle description holds, which gives the shift of the single-particle energies as well as the shift of the bound state energies. Also the modification of the scattering phase shifts due to medium effects is calculated. These quantities then can be considered as input to calculate the EoS at higher densities. Different approximations for the quasiparticle shifts have been considered here.

A description of nuclear matter at low densities should also include the formation of other bound states, that is, nuclei in the ground state and excited states. Of particular interest is the α particle as the binding energy per nucleon is rather high so that it will be abundant in nuclear matter. The inclusion of both three and four-particle correlations in nuclear matter allows not only to describe the abundances of t, h, α but also their influence on the EoS and phase transitions.

An interesting effect is the dissolution of all bound states with increasing density. Because of the Pauli blocking, the bound state energy is shifted and merges with the continuum at the so-called Mott density, which depends on the temperature and the c.o.m. momentum of the bound state. This effect has the consequence that near the saturation density bound states are dissolved in nuclear matter, and the single-nucleon quasiparticle approach becomes valid. Below the Mott density, bound states are of relevance in calculating the properties of nuclear matter. In general, we find approximations that interpolate between the low-density virial expansion and the Fermi liquid approach at higher densities.

Cluster formation is essential for the symmetry energy in the low-density region. A quasiparticle approach fails to give correct results, which are determined by the formation of bound states such as α particles. A challenging issue is also the microscopic approach to α matter, which should include the simultaneous treatment of the single-nucleon states and other correlations.

Below a critical temperature, depending on baryon density and asymmetry, the deuterons will form a Bose–Einstein condensate. With increasing density, the bound states are dissolved, and we observe the cross-over from BEC to Cooper pairing. Pairing is also of relevance in finite nuclei.

Quantum condensates in nuclear matter are treated beyond the mean-field approximation, with the inclusion of cluster formation. The occurrence of a separate binding pole in the four-particle propagator in nuclear matter is investigated with respect to the formation of a condensate of α-like particles (quartetting), which is dependent on temperature and density.

Because of interaction effects and Pauli blocking, the formation of an α-like condensate is limited to the low-density region. Consequences for finite nuclei are considered. In particular, excitations of n-α self-conjugate nuclei ($Z = 2n$, $N = 2n$) near the n-α breakup threshold are candidates for quartetting, for example, ^{8}Be, ^{12}C, and ^{16}O. Exploratory calculations are performed for the density dependence of the α condensate fraction at zero temperature to address the suppression of the four-particle condensate below nuclear matter saturation density.

The microscopic approach to the EoS anticipates a homogeneous system in equilibrium, described by the grand canonical ensemble. However, thermodynamic stability with respect to phase separation has to be shown. For a special density region depending on the asymmetry, nuclear matter becomes instable below the critical temperature for the gas–liquid like phase transition. The spinodale marks the region where spontaneous phase separation occurs. The EoS, calculated for a homogeneous system, contradicts in that region the stability conditions. If the Coulomb interaction and a background of neutralizing electrons is taken into account, instead of phase separation, droplet formation and other structures will characterize the region of two-phase instability.

In nuclear reactions, particularly in heavy ion collisions, we have inhomogeneous matter in nonequilibrium, and the nuclear matter EoS serves only as an approximation to characterize a state in local equilibrium, but changing with space and time. Nevertheless, this local equilibrium may be considered as a prerequisite to describe the formation of correlations and bound states. Simple approximations such as the freeze-out concept or the coalescence model are presently used to characterize the cluster formation in heavy ion collisions.

Important consequences are also expected for nonequilibrium processes occurring in astrophysical objects. Cluster formation is of importance in other nonequilibrium processes and will determine the mean free path for different particles. An interesting application is the neutrino transport in supernova collapses. The formation of clusters is of interest in the early universe, if inhomogeneous distribution of matter as a consequence of phase separation is considered [52].

Acknowledgment

The author is indebted for stimulating discussions and fruitful collaborations with a lot of nuclear physicists, in particular M. Beyer, D. Blaschke, C. Fuchs, Y. Funaki, H. Horiuchi, T. Klähn, J. Natowitz, S. Schlomo, P. Schuck, A. Sedrakian, K. Sumiyoshi, A. Tohsaki, S. Typel, A. Wierling, H. Wolter, and T. Yamada.

References

1. M. Baldo (ed.), *Nuclear Methods and the Nuclear Equation of State* (World Scientific, Singapore, 1999)
2. T. Klähn et al., Phys. Rev. C **74**, 035802 (2006)
3. A.L. Fetter, J.D. Walecka, *Quantum Theory of Many-Particle Systems* (McGraw-Hill, New York, 1971); A.A. Abrikosov, L.P. Gorkov, I.E. Dzyaloshinski, *Methods of Quantum Field Theory in Statistical Mechanics* (Dover, New York, 1975)
4. G. Röpke, L. Münchow, H. Schulz, Nucl. Phys. A **379** , 536 (1982); G. Röpke, M. Schmidt, L. Münchow, H. Schulz, Phys. Lett. B **110**, 21 (1982)
5. G. Röpke, T. Seifert, H. Stolz, R. Zimmermann, Phys. Stat. Sol. B **100**, 215 (1980); G. Röpke, M. Schmidt, L. Münchow, H. Schulz, Nucl. Phys. A **399**, 587 (1983); G. Röpke, in *Aggregation Phenomena in Complex Systems*, ed. by J. Schmelzer et al., Chaps. 4, 12 (Wiley-VCH, New York, 1999); Dukelsky, G. Röpke, P. Schuck, Nucl. Phys. A **628**, 17 (1998)
6. M. Schmidt, G. Röpke, H. Schulz, Ann. Phys. (NY) **202**, 57 (1990)
7. M. Beyer, W. Schadow, C. Kuhrts, G. Röpke, Phys. Rev. C **60**, 034004 (1999); M. Beyer, S.A. Sofianos, C. Kuhrts, G. Röpke, P. Schuck, Phys. Letters B **488**, 247 (2000)
8. H. Stein, A. Schnell, T. Alm, G. Röpke, Z. Phys. A **351**, 295 (1995)
9. G. Röpke, A. Schnell, P. Schuck, P. Nozières, Phys. Rev. Lett. **80**, 3177 (1998)
10. G. Röpke, A. Schnell, P. Schuck, U. Lombardo, Phys. Rev. C **61**, 024306 (2000)
11. A. Tohsaki, H. Horiuchi, P. Schuck, G. Röpke, Phys. Rev. Lett. **87**, 192501 (2001)
12. H. Shen, H. Toki, K. Oyamatsu, K. Sumiyoshi, Progr. Theor. Phys. **100**, 1013 (1998); N. Ohnishi, K. Kotake, S. Yamada, Astrophys. J. **667**, 375 (2006)
13. J. Haidenbauer, W. Plessas, Phys. Rev. C **30**, 1822 (1984); L. Mathelitsch, W. Plessas, W. Schweiger, Phys. Rev. C **36**, 65 (1986)
14. C.J. Horowitz, A. Schwenk, Nucl. Phys. A **776**, 55 (2006)
15. E. O'Connor, D. Gazit, C.J. Horowitz, A. Schwenk, N. Barnea, Phys. Rev. C **75**, 055803 (2007)
16. Y. Yamaguchi, Phys. Rev. **95**, 1628 (1954); J.F. Berger, M. Girod, D. Gogny, Comp. Phys. Comm **63**, 365 (1991); T.R. Mongan, Phys. Rev. **178**, 1597 (1969)
17. J. Margueron, E. van Dalen, C. Fuchs, Phys. Rev. C **76**, 034309 (2007) [arXiv:0707.0354 [nucl-th]]
18. J.M. Lattimer, F.D. Swesty, Nucl. Phys. A **535**, 331 (1991)
19. S. Typel, Phys. Rev. C **71**, 064301 (2005)
20. G. Röpke, M. Schmidt, H. Schulz, Nucl. Phys. A **424**, 594 (1984)
21. G. Röpke, A. Grigo, K. Sumiyoshi, Hong Shen, Phys. Part. Nucl. Lett. **2**, 275 (2005)
22. G. Audi, A.H. Wapstra, Nucl. Phys. A **565**, 1 (1993)
23. G. Röpke, A. Grigo, K. Sumiyoshi, Hong Shen, in *Superdense QCD Matter and Compact Stars*, ed. by D. Blaschke, A. Sedrakian. NATO Science Series (Springer, Dordrecht, 2006), pp. 75–91
24. G. Röpke, in *Clusters in Nuclear Matter*, ed. by S. Hernandez, J.W. Clark. Condensed Matter Theories, vol 16 (Nova Sciences Publ., New York, 2001)
25. S. Kowalski et al., Phys. Rev. C **75**, 014601 (2007)
26. G. Röpke, Ann. Physik (Leipzig) **3**, 145 (1994); R. Haussmann, Z. Physik B **91**, 291 (1993)

27. P. Nozières, S. Schmitt-Rink, J. Low Temp. Phys. **59**, 159 (1985)
28. A. Schnell, G. Röpke, P. Schuck, Phys. Rev. Lett. **83**, 1926 (1999)
29. M. Randeria, Varenna Lectures 1997, cond-mat/9710223
30. W.H. Dickhoff, C. Barbieri, Prog. Part. Nucl. Phys. **52**, 377 (2004); A. Schnell, Ph.D. Thesis, Rostock, 1996
31. T. Alm, B.L. Friman, G. Röpke, H. Schulz, Nucl. Phys. A **551**, 45 (1993)
32. O. Penrose, L. Onsager, Phys. Rev. **104**, 576 (1956)
33. M.T. Johnson, J.W. Clark, Kinam **2**, 3 (1980) (PDF available at Faculty web page of J.W. Clark at http://wuphys.wustl.edu); see also J.W. Clark, T.P. Wang, Ann. Phys. (N.Y.) **40**, 127 (1966) and G.P. Mueller, J.W. Clark, Nucl. Phys. A **155**, 561 (1970)
34. K.A. Gernoth, M.L. Ristig, T. Lindenau, Int. J. Mod. Phys. B **21**, 2157 (2007); G. Senger, M.L. Ristig, C.E. Campbell, J.W. Clark, Ann. Phys. (N.Y.) **218**, 160 (1992); R. Pantfoerder, T. Lindenau, M.L. Ristig, J. Low Temp. Phys. **108**, 245 (1997)
35. S. Ali, A.R. Bodmer, Nucl. Phys. A **80**, 99 (1966)
36. G. Röpke, P. Schuck, Mod. Phys. Lett. A **21**, 2513 (2006)
37. Z.F. Shehadeh et al., Int. J. Mod. Phys. B **21**, 2429 (2007); M.N.A. Abdullah et al., Nucl. Phys. A **775**, 1 (2006)
38. A. Tohsaki, H. Horiuchi, P. Schuck, G. Röpke, Phys. Rev. Lett. **87**, 192501 (2001)
39. Y. Funaki, H. Horiuchi, G. Röpke, P. Schuck, A. Tohsaki, T. Yamada (in preparation)
40. Y. Funaki, T. Yamada, H. Horiuchi, G. Röpke, P. Schuck, A. Tohsaki, Phys. Rev. Lett. 101, 082502 (2008)
41. Y. Fujiwara, H. Horiuchi, K. Ikeda, M. Kamimura, K. Kato, Y. Suzuki, E. Uegaki, Prog. Theor. Phys. Suppl. **68**, 29 (1980)
42. E. Uegaki, S. Okabe, Y. Abe, H. Tanaka, Prog. Theor. Phys. **57**, 1262 (1977); E. Uegaki, Y. Abe, S. Okabe, H. Tanaka, Prog. Theor. Phys. **59**, 1031 (1978); **62**, 1621 (1979)
43. Y. Fukushima, M. Kamimura, *Proc. Int. Conf. on Nuclear Structure*, Tokyo, 1977, ed. by T. Marumori (Suppl. of J. Phys. Soc. Japan, Vol. 44, 1978), p. 225; M. Kamimura, Nucl. Phys. A **351**, 456 (1981)
44. P. Navrátil, J.P. Vary, B.R. Barrett, Phys. Rev. Lett. **84**, 5728 (2000); P. Navrátil, J.P. Vary, B.R. Barrett, Phys. Rev. C **62**, 054311 (2000); B.R. Barrett, B. Mihaila, S.C. Pieper, R.B. Wiringa, Nucl. Phys. News, **13**, 17 (2003)
45. H. Horiuchi, Prog. Theor. Phys. **51**, 1266 (1974); **53**, 447 (1975)
46. Y. Funaki, H. Horiuchi, A. Tohsaki, P. Schuck, G. Röpke, Prog. Theor. Phys. **108**, 297 (2002)
47. M. Chernykh, H. Feldmeier, T. Neff, P. von Neumann-Cosel, A. Richter, Phys. Rev. Lett. **98**, 032501 (2007)
48. H. Matsumura, Y. Suzuki, Nucl. Phys. A **739**, 238 (2004)
49. T. Yamada, P. Schuck, Eur. Phys. J. A **26**, 185 (2005)
50. T. Wakasa et al., Phys Lett B **653**, 173 (2007)
51. Y. Funaki, H. Horiuchi, G. Röpke, P. Schuck, A. Tohsaki, T. Yamada, Nucl. Phys. News **17**(04), 11 (2007)
52. G. Röpke, Phys. Lett. B **185**, 281 (1987)

7

BEC–BCS Crossover in Strongly Interacting Matter

Daniel Zablocki, David Blaschke, and Gerd Röpke

Abstract. A quantum field theoretical approach to the thermodynamics of dense Fermi systems is developed for the description of the formation and the dissolution of quantum condensates and bound states in dependence of temperature and density. As a model system, we study the chiral and superconducting phase transitions in two-flavor quark matter within the NJL model and their interrelation with the formation of quark–antiquark and diquark bound states. The phase diagram of quark matter is evaluated as a function of the diquark coupling strength, and a coexistence region of chiral symmetry breaking and color superconductivity is obtained at very strong coupling. The crossover between Bose–Einstein condensation of diquark bound states and condensation of diquark resonances (Cooper pairs) in the continuum is discussed as a Mott effect. This effect consists in the transition of bound states into the continuum of scattering states under the influence of compression and heating. We explain the physics of the Mott transition, with special emphasis on the role of the Pauli principle for the case of the pion in quark matter.

7.1 Introduction

Key issues of modern physics of dense matter are concepts explaining the phenomena related to the appearance of quantum condensates in dense Fermi systems. Two regimes are well-known: the Bose–Einstein condensation (BEC) of bound states with an even number of fermions and the condensation of bosonic correlations (e.g., Cooper pairs) in the continuum of unbound states according to the Bardeen–Cooper–Schrieffer (BCS) theory. While the former mechanism concerns states that are well-localized in coordinate space as they occur for strong enough attractive coupling, the latter mechanism applies to states that are correlated within a shell of the order of the energy gap Δ around the Fermi sphere in momentum space but delocalized in coordinate space. The transition between both regimes is called BEC–BCS crossover. Recently, this transition regime became accessible to laboratory experiments with ultracold gases of fermionic atoms coupled via Feshbach resonances, with a strength tunable by applying external magnetic fields, see Fig. 7.1. After the

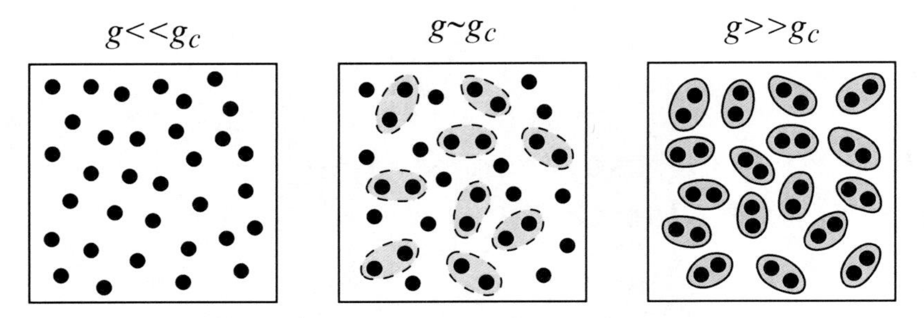

Fig. 7.1. Illustration of the transition from BCS pairing with delocalized wave functions to BEC of bound states, well-localized in coordinate space, from [1]

preparation of fermionic dimers in 2003, now also the BEC [2, 3] and the superfluidity of these dimers have been observed [4, 5]. The BEC–BCS crossover is physically related [6] to the bound state dissociation or Mott–Anderson delocalization transition [7], where the modification of the effective coupling strength is caused by electronic screening and/or Pauli blocking effects. It is thus a very general effect expected to occur in a wide variety of dense Fermi systems with attractive interactions [8] such as electron–hole systems in solid state physics [9], electron–proton systems in the interior of giant planets [10], deuterons in nuclear matter [11–13], or diquarks in quark matter [14–17]. The BEC–BCS crossover transition in quark matter takes place when forming or dissolving hadrons at the Mott density when the temperatures are low enough for condensation of the bosonic correlations. It is of particular theoretical interest due to the additional relativistic regime when the binding energy compensates for the mass of constituents, and the correlations may therefore reach a massless, ultrarelativistic limit [18–20].

A systematic treatment of these effects is possible within the path integral formulation for finite-temperature quantum field theories. This approach is rather general as it is relativistic and is especially suited to take into account the effects of spontaneous symmetry breaking, manifest, for example, in the chiral symmetry breaking and color superconductivity transitions in quark matter or in the superfluid and liquid–gas phase transitions in nuclear matter. Within this contribution, we present the basics of this approach on the example of a model field theory of the Nambu–Jona–Lasinio type for relativistic, strongly interacting Fermi systems. These investigations are also motivated by the analogies of the strongly coupled quark–gluon plasma (sQGP) at Relativistic Heavy Ion Collider (RHIC) in Brookhaven [21], with the experiments on BEC of atoms in traps. Furthermore, qualitative insights into possible effects observable in the upcoming CBM experiment at FAIR Darmstadt as well as from neutron stars with quark matter interiors could be derived along the lines of this approach.

7.2 Quark Matter

7.2.1 Partition Function and Model Lagrangian

As a generic model system for the description of hot, dense Fermi-systems with strong, short-range interactions, we consider quark matter described by a model Lagrangian with four-fermion coupling. The key quantity for the derivation of thermodynamic properties is the partition function $\mathcal{Z}$ from which all thermodynamic quantities can be derived. It is given as a path integral, which in the imaginary time formalism ($t = -\mathrm{i}\tau$) can be expressed as [22]

$$\mathcal{Z} = \int \mathcal{D}(\mathrm{i}q^\dagger)\mathcal{D}(q)\, \mathrm{e}^{\int^\beta \mathrm{d}^4 x\, (\mathcal{L} - \mu q^\dagger q)}, \tag{7.1}$$

where the chemical potential μ is introduced as a Lagrange multiplier for assuring conservation of baryon number as a conserved charge carried by the quarks. The notation $\int^\beta \mathrm{d}^4x$ is shorthand for $\int_0^\beta \mathrm{d}\tau \int \mathrm{d}^3x$, where $\beta = 1/T$ is the inverse temperature. The quark matter is described by a Dirac Lagrangian with internal degrees of freedom ($N_\mathrm{f} = 2$ flavors , $N_\mathrm{c} = 3$ colors), with a current–current-type four-fermion interaction inspired by one-gluon exchange

$$\mathcal{L} = \bar{q}(\mathrm{i}\gamma_\mu \partial^\mu - m_0)q - \frac{g^2}{2} \sum_{a=1}^{8} \bar{q}\frac{\lambda^a}{2}\gamma_\mu q\, \bar{q}\frac{\lambda^a}{2}\gamma^\mu q, \tag{7.2}$$

where λ^a are the Gell–Mann matrices for color $SU(3)$. After Fierz transformation of the interaction, we select the scalar diquark channel and the scalar, pseudoscalar, and vector meson channels so that our model Lagrangian assumes the form

$$\mathcal{L} = \mathcal{L}_0 + \mathcal{L}_{qq} + \mathcal{L}_{q\bar{q}}, \tag{7.3}$$

where the different terms are given by

$$\mathcal{L}_0 = \bar{q}(\mathrm{i}\gamma_\mu \partial^\mu - m_0)q, \tag{7.4}$$

$$\mathcal{L}_{q\bar{q}} = G_\mathrm{S}\left[(\bar{q}q)^2 + (\bar{q}\mathrm{i}\gamma_5 \tau q)^2\right], \tag{7.5}$$

$$\mathcal{L}_{qq} = G_\mathrm{D}\left\{\bar{q}\left[\mathrm{i}\gamma_5 C\tau_2\lambda_2\right]\bar{q}^\mathrm{T}\right\}\left\{q^\mathrm{T}\left[\mathrm{i}C\gamma_5\tau_2\lambda_2\right]q\right\}, \tag{7.6}$$

where γ_ν are the Dirac matrices, τ_i are $SU(2)$ flavor matrices and $C = \mathrm{i}\gamma^2\gamma^0$ is the charge conjugation matrix. G_S and G_D are the coupling strengths corresponding to the different channels, see [23] for a recent review. For the numerical analysis we adopt parameters from [24], that is, $\Lambda = 629.5\,\mathrm{MeV}$, $m_0 = 5.3\,\mathrm{MeV}$, and $G_\mathrm{S}\Lambda^2 = 2.18$, and consider G_D as a free parameter of the model.

A general method to deal with four-fermion interactions in the path integral approach starts with the Hubbard–Stratonovich transformation [25] of the partition function to its equivalent form in terms of collective bosonic fields, which is more suitable to deal with nonperturbative effects such as the occurrence of order parameters (mean fields) related to phase transitions in the system. The correlations beyond the mean fields are mesons corresponding to plasmon (particle–hole) excitations and diquarks corresponding to pair fluctuations (two-particle bound and scattering states) in the dense fermion system.

7.2.2 Hubbard–Stratonovich Transformation: Bosonization

The Hubbard–Stratonovich transformation is a two-step procedure, which consists of (1) linearization of the four-fermion interaction terms by introducing bosonic auxiliary fields in the appropriate channels and (2) integrating out the fermions analytically.

We introduce the Hubbard–Stratonovich auxiliary fields $\Delta(\tau,x)$, $\Delta^*(\tau,x)$, $\pi(\tau,x)$, and $\sigma(\tau,x)$ so that the partition function of the system becomes

$$\mathcal{Z} = \int \mathcal{D}\Delta^*\mathcal{D}\Delta\mathcal{D}\sigma\mathcal{D}\pi \left\{ \exp\left\{ -\int^{\beta} \mathrm{d}^4x \left[\frac{\sigma^2+\pi^2}{4G_{\mathrm{S}}} + \frac{|\Delta|^2}{4G_{\mathrm{D}}} \right] \right\} \int [\mathrm{d}q]\,[\mathrm{d}\bar{q}] \right.$$
$$\times \exp\left\{ \int^{\beta} \mathrm{d}^4x \, (\bar{q}(\mathrm{i}\gamma_\mu\partial^\mu + \mu\gamma_0 - m_0)q - \bar{q}(\sigma + \mathrm{i}\gamma_5\tau\cdot\pi)q \right.$$
$$\left.\left. -\frac{\Delta^*}{2}q^{\mathrm{T}}Rq - \frac{\Delta}{2}\bar{q}\tilde{R}\bar{q}^{\mathrm{T}} \Big) \right\} \right\}, \tag{7.7}$$

where $R = \mathrm{i}C\gamma_5\otimes\tau_2\otimes\lambda_2$, $\tilde{R} = \mathrm{i}\gamma_5 C\otimes\tau_2\otimes\lambda_2$. By introducing Nambu–Gorkov spinors

$$\Psi \equiv \frac{1}{\sqrt{2}}\begin{pmatrix} q \\ q^{\mathrm{c}} \end{pmatrix}, \quad \bar{\Psi} \equiv \frac{1}{\sqrt{2}}\left(\bar{q}\bar{q}^{\mathrm{c}}\right) \tag{7.8}$$

with $q^{\mathrm{c}}(x) \equiv C\bar{q}^{\mathrm{T}}(x)$, the Lagrangian takes the bilinear form

$$\mathcal{L} = \bar{\Psi}\begin{pmatrix} \mathrm{i}\gamma_\mu\partial^\mu + \mu\gamma_0 - \hat{m} - \mathrm{i}\gamma_5\tau\cdot\pi & \mathrm{i}\Delta\gamma_5\tau_2\lambda_2 \\ \mathrm{i}\Delta^*\gamma_5\tau_2\lambda_2 & i\gamma_\mu\partial^\mu - \mu\gamma_0 - \hat{m} - i\gamma_5\tau\cdot\pi \end{pmatrix}\Psi \tag{7.9}$$

with $\hat{m} = m_0 + \sigma$. Hence the partition function becomes a Gaussian path integral in the bispinor fields, which can be evaluated and yields the fermion determinant

$$\mathcal{Z} = \int \mathcal{D}\Delta^*\mathcal{D}\Delta\mathcal{D}\sigma\mathcal{D}\pi \, \exp\left\{ -\int^{\beta} \mathrm{d}^4x \frac{\sigma^2+\pi^2}{4G_{\mathrm{S}}} + \frac{|\Delta|^2}{4G_{\mathrm{D}}} \right\}$$
$$\times \int \mathcal{D}\bar{\Psi}\mathcal{D}\Psi \, \exp\left\{ \int^{\beta} \mathrm{d}^4x \bar{\Psi}\left[S^{-1}\right]\Psi \right\} \tag{7.10}$$
$$= \int \mathcal{D}\Delta^*\mathcal{D}\Delta\mathcal{D}\sigma\mathcal{D}\pi \, \exp\left\{ -\int^{\beta} \mathrm{d}^4x \frac{\sigma^2+\pi^2}{4G_{\mathrm{S}}} + \frac{|\Delta|^2}{4G_{\mathrm{D}}} \right\} \mathrm{Det}[S^{-1}], \tag{7.11}$$

where the inverse bispinor propagator is a matrix in Nambu–Gorkov-, Dirac-, color-, and flavor-space, which after Fourier transformation reads

$$S^{-1} = \begin{pmatrix} (\mathrm{i}\omega_n + \mu)\gamma_0 - \hat{m} - \mathrm{i}\gamma\mathbf{p} - \mathrm{i}\gamma_5\tau \cdot \pi & \mathrm{i}\Delta\gamma_5\tau_2\lambda_2 \\ \mathrm{i}\Delta^*\gamma_5\tau_2\lambda_2 & (\mathrm{i}\omega_n - \mu)\gamma_0 - \hat{m} - \mathrm{i}\gamma\mathbf{p} + \mathrm{i}\gamma_5\tau \cdot \pi. \end{pmatrix}. \quad (7.12)$$

So far we could derive with (7.11) a very compact, bosonized form of the quark matter partition function (7.1), which is an exact transformation of (7.1), now formulated in terms of collective, bosonic fields. As we demonstrate in the following, this form is suitable as it allows to obtain nonperturbative results already in the lowest orders with respect to an expansion around the stationary values of these fields. In performing this expansion, we may factorize the partition function into mean field (MF), Gaussian fluctuation (Gauss), and residual (res) contributions

$$Z(\mu, T) \equiv \mathrm{e}^{-\beta\Omega(\mu,T)} = Z_{\mathrm{MF}}(\mu, T) Z_{\mathrm{Gauss}}(\mu, T) Z_{\mathrm{res}}(\mu, T).$$

In the following, we discuss the physical content of these approximations.

7.2.3 Mean-Field Approximation: Order Parameters

After the evaluation of the traces in the internal spaces and the sum over the Matsubara frequencies, one gets

$$\begin{aligned} \Omega_{\mathrm{MF}} &= -\frac{1}{\beta V} \ln Z_{\mathrm{MF}} = \frac{(m - m_0)^2}{4G_{\mathrm{S}}} + \frac{|\Delta|^2}{4G_{\mathrm{D}}} - \frac{1}{\beta V} \mathrm{Tr}\left(\ln \beta S_{\mathrm{MF}}^{-1}\right), \\ &= \frac{(m - m_0)^2}{4G_{\mathrm{S}}} + \frac{|\Delta|^2}{4G_{\mathrm{D}}} - 4\int \frac{\mathrm{d}^3 p}{(2\pi)^3} \Big[E_{\mathbf{p}}^+ + E_{\mathbf{p}}^- + E_{\mathbf{p}} + 2T \ln(1 + \mathrm{e}^{-\beta E_{\mathbf{p}}^+}) \\ &\quad + 2T \ln(1 + \mathrm{e}^{-\beta E_{\mathbf{p}}^-}) + T \ln(1 + \mathrm{e}^{-\beta \xi_{\mathbf{p}}^+}) + T \ln(1 + \mathrm{e}^{-\beta \xi_{\mathbf{p}}^-})\Big], \end{aligned} \quad (7.13)$$

where we have defined the particle dispersion relation $E_{\mathbf{p}}^{\pm} = \sqrt{\left(\xi_{\mathbf{p}}^{\pm}\right)^2 + \Delta^2}$ with $\xi_{\mathbf{p}}^{\pm} = E_{\mathbf{p}} \pm \mu$, $E_{\mathbf{p}} = \sqrt{m^2 + \mathbf{p}^2}$. The $\Delta \neq 0$ dispersion law is associated to the red and green quarks ($E_{\mathbf{p}}^-$) and antiquarks ($E_{\mathbf{p}}^+$), whereas the ungapped blue quarks (antiquarks) have the dispersion $\xi_{\mathbf{p}}^-$ ($\xi_{\mathbf{p}}^+$). In thermodynamical equilibrium, the mean field values satisfy the stationarity condition of the minimal thermodynamical potential, that is,

$$\frac{\partial \Omega_{\mathrm{MF}}}{\partial \sigma_{\mathrm{MF}}} = \frac{\partial \Omega_{\mathrm{MF}}}{\partial \pi_{\mathrm{MF}}} = \frac{\partial \Omega_{\mathrm{MF}}}{\partial \Delta_{\mathrm{MF}}} = 0, \quad (7.14)$$

equivalent to the fulfillment of the gap equations $\sigma_{\mathrm{MF}} = -4G_{\mathrm{S}}\mathrm{Tr}\left(S_{\mathrm{MF}}\right) \equiv m - m_0$, $\boldsymbol{\pi}_{\mathrm{MF}} = -4\mathrm{i}G_{\mathrm{S}}\mathrm{Tr}\left(\gamma_5\boldsymbol{\tau} S_{\mathrm{MF}}\right) = 0$ and $\Delta_{\mathrm{MF}} = 4G_{\mathrm{D}}\mathrm{Tr}\left(\gamma_5\tau_2\lambda_2 S_{\mathrm{MF}}\right) = \Delta$, together with the stability criterion that the determinant of the curvature matrix formed by the second derivatives is positive. From (7.14) with (7.13), we obtain the gap equations for the order parameters m and Δ, which have to be solved self-consistently,

$$m - m_0 = 8G_S\, m \int \frac{d^3p}{(2\pi)^3} \frac{1}{E_{\mathbf{p}}} \Big\{ \left[1 - 2n_F(E^-_{\mathbf{p}})\right] \frac{\xi^-_{\mathbf{p}}}{E^-_{\mathbf{p}}} + \left[1 - 2n_F(E^+_{\mathbf{p}})\right] \frac{\xi^+_{\mathbf{p}}}{E^+_{\mathbf{p}}} + n_F(-\xi^+_{\mathbf{p}}) - n_F(\xi^-_{\mathbf{p}}) \Big\}, \tag{7.15}$$

$$\Delta = 8G_D \int \frac{d^3p}{(2\pi)^3} \left[\frac{1 - 2n_F(E^-_{\mathbf{p}})}{E^-_{\mathbf{p}}} + \frac{1 - 2n_F(E^+_{\mathbf{p}})}{E^+_{\mathbf{p}}} \right], \tag{7.16}$$

with the Fermi distribution function $n_F(E) = (1 + e^{\beta E})^{-1}$. For zero temperature, the gap equations take the simple form

$$m - m_0 = 8G_S m \int \frac{d^3p}{(2\pi)^3} \frac{1}{E_{\mathbf{p}}} \left[\frac{\xi^-_{\mathbf{p}}}{E^-_{\mathbf{p}}} + \frac{\xi^+_{\mathbf{p}}}{E^+_{\mathbf{p}}} + \Theta(\xi^-_{\mathbf{p}}) \right], \tag{7.17}$$

$$\Delta = 8G_D\, \Delta \int \frac{d^3p}{(2\pi)^3} \left[\frac{1}{E^-_{\mathbf{p}}} + \frac{1}{E^+_{\mathbf{p}}} \right]. \tag{7.18}$$

Solutions of the gap equations for the dynamically generated quark mass m and for the diquark pairing gap Δ at $T = 0$ as a function of the chemical potential are shown in Fig. 7.2. From the knowledge of the order parameters as functions of the thermodynamical variables (T, μ), we can deduce the phase diagram of Fig. 7.3.

7.2.4 Phase Diagram

From the solutions of the gap equations for the order parameters in dependence of the thermodynamical variables T and μ, we have constructed the

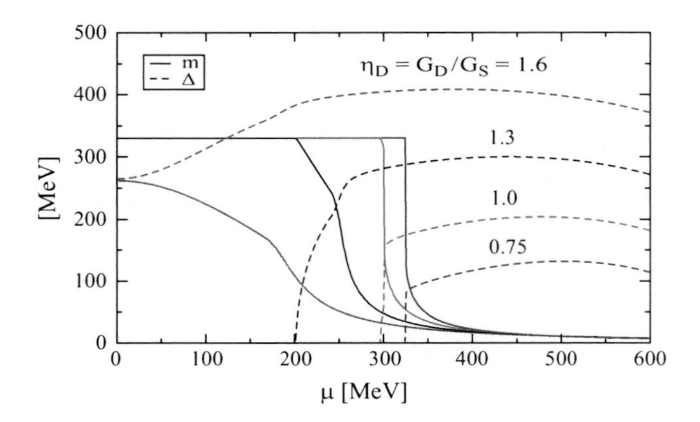

Fig. 7.2. Order parameters for chiral symmetry breaking (m, *full lines*) and color superconductivity (Δ, *dashed lines*) at $T = 0$ for different values of the diquark coupling η_D. First order phase transitions turn to second order or even crossover when η_D is increased. For details, see text

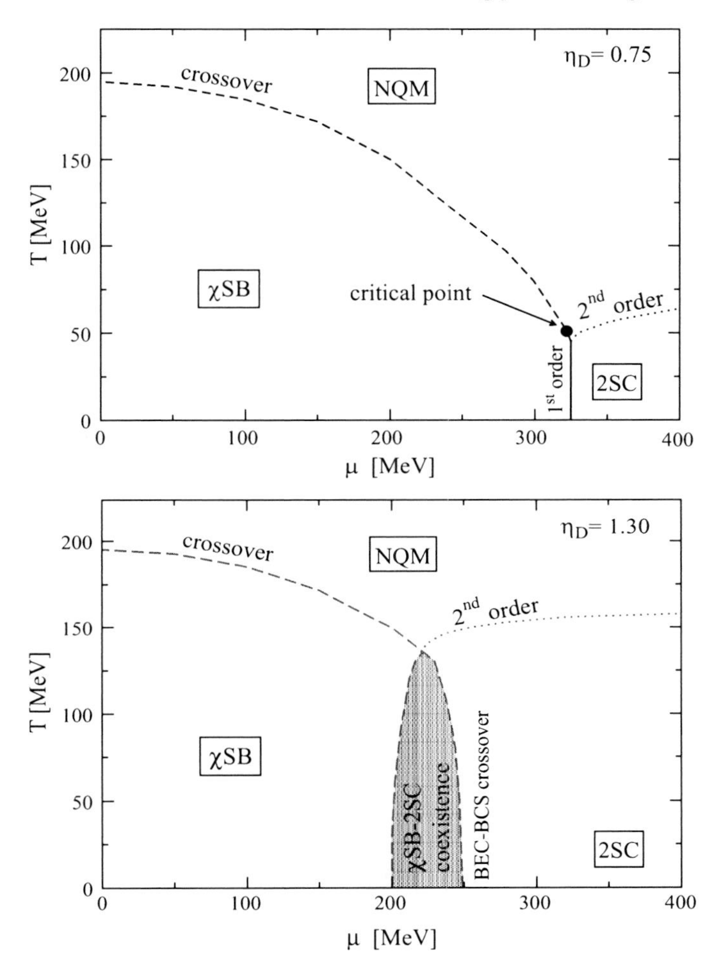

Fig. 7.3. Phase diagram of two-flavor quark matter with critical lines for chiral symmetry breaking and color superconductivity for different values of the diquark coupling strength η_D. *Upper panel:* For rather small coupling $\eta_D = 0.75$, the phase diagram features a first order phase transition (*solid line*) for the quark mass m as well as the diquark gap Δ for low temperatures, which changes into crossover (m, *dashed line*) and second order (Δ, *dotted line*) for higher temperatures. We also find a critical endpoint for first order phase transitions. *Lower panel:* For larger coupling $\eta_D = 1.30$, we do not find first order phase transitions anymore. For increasing temperatures, the diquark gap undergoes a second-order phase transition (*dotted line*), whereas all other phase transitions turn out to be crossover. The BEC–BCS crossover occurs when the chiral transition (coincident with the Mott transition for mesonic and diquark bound states) occurs inside the 2SC phase. It is characterized by the coexistence of diquark condensation with chiral symmetry breaking

phase diagram of the present quark matter model in the $T - \mu$ plane, see Fig. 7.3. The two order parameters allow to distinguish four phases:

- $\Delta = 0$, $m \sim m_0$: normal phase (NQM)
- $\Delta \neq 0$, $m \sim m_0$: color superconductor (2SC)
- $\Delta = 0$, $m \gg m_0$: chiral symmetry broken phase (χSB)
- $\Delta \neq 0$, $m \gg m_0$: coexistence of χSB and 2SC (BEC phase)

Order parameters are indicators of phase transitions. The phase transitions can be classified according to their order, depending on the behavior of the order parameters with the change of thermodynamic variables:

- First order: order parameter jumps, as in the case of the χSB $\rightarrow$ 2SC phase transition at not too large coupling.
- Second order: order parameter turns continuously to zero. The 2SC $\rightarrow$ NQM transition with increasing T is always second order. The χSB $\rightarrow$ 2SC transition turns from first to second order for strong enough coupling.
- Crossover: the order parameter changes continuously, but does not go to zero and also has no jumps. An example is the χSB $\rightarrow$ NQM transition at temperatures above the critical endpoint (CP), where it goes over to the line of first order transitions in the $T - \mu$ plane. The identification of the CP is a key issue for experimental research and thus for the verification of QCD models. It is suggested that verifiable signatures (change of the fluctuation spectrum, latent heat or not) are related with it. For strong coupling, the CP moves to lower T and finally to $T = 0$ (for $\eta_D > 1.3$, the chiral transition is always crossover).

Increasing the diquark coupling η_D leads to an increase of the diquark gap and therefore a rise in the critical temperature for the second-order transition to a normal quark matter phase. It shifts also the border between color superconductivity (2SC) and chiral symmetry broken phase (χSB) to lower values of the chemical potential. For very strong coupling $\eta_D \sim 1$, a coexistence region develops, where both order parameters are simultaneously nonvanishing. Under these conditions, the phase border is not of first order and therefore no critical endpoint can be identified. As we are going to explain in the next section, in the χSB phase, pion and diquark bound states can exist. At the chiral symmetry restoration transition, they merge the continuum of unbound states and turn into (resonant) scattering states. When this Mott transition occurs within the 2SC phase (characterized by a nonvanishing diquark condensate), we speak of a BEC–BCS crossover: the condensation of diquark bound states (BEC) turns into a condensation of resonances, called Cooper pairs (BCS).

For lower coupling, the critical point occurs and is shown as a colored dot in the phase diagram of Fig. 7.3.

In the next section, we turn towards the interesting question about the quasiparticle excitations in these phases. To this end, we expand the action functional in the partition function up to quadratic (Gaussian) order in the mesonic fields and arrive at a tractable approximation for the bosonized quark matter model (7.11).

7.2.5 Gaussian Fluctuations: Bound and Scattering States

Let us expand now the mesonic fields around their mean field values. In this article, we focus on fluctuations in the mesonic channels, where the pion and the sigma meson will emerge as quasiparticle degrees of freedom. On the example of the pion, we explain the physics of the Mott transition. As discussed in the previous section, the phenomenon of the BEC–BCS crossover in the 2SC phase is due to the Mott transition for diquarks. The detailed investigation of the quantized diquark fluctuations, which are also a prerequisite of the formation of baryons, will be given elsewhere [20, 26, 27, 27a]. As we already noticed, the pion does not contribute to the mean field, and we need to introduce only the sigma-field fluctuations as $\sigma \to \sigma_{\rm MF} + \sigma$. Hence it is possible to decompose the inverse propagator S^{-1} into a mean field part and a fluctuation part $S^{-1} = S^{-1}_{\rm MF} + \Sigma$, where the matrix Σ is defined as

$$\Sigma \equiv \begin{pmatrix} -\sigma - {\rm i}\gamma_5 \boldsymbol{\tau}\cdot\boldsymbol{\pi} & 0 \\ 0 & -\sigma - {\rm i}\gamma_5 \boldsymbol{\tau}^t\cdot\boldsymbol{\pi} \end{pmatrix}. \tag{7.19}$$

In the Gaussian approximation, the fermion determinant becomes

$$\frac{{\rm Det}\left[S^{-1}\right]|_{\rm Gauss}}{{\rm Det}\left[S^{-1}_{\rm MF}\right]} = \exp\Big\{ -\frac{1}{2}\int\frac{{\rm d}^4 q}{(2\pi)^4} \times {\rm Tr}\left[S_{\rm MF}(p)\Sigma(q)S_{\rm MF}(p+q)\Sigma(q)\right]\Big\}. \tag{7.20}$$

The propagator $S_{\rm MF}$ is obtained from (7.12) by the matrix inversion

$$S_{\rm MF} \equiv \begin{pmatrix} \mathbf{G}^+ & \mathbf{F}^- \\ \mathbf{F}^+ & \mathbf{G}^- \end{pmatrix}, \tag{7.21}$$

with the matrix elements

$$\mathbf{G}^\pm_p = \sum_{s_p}\sum_{t_p}\frac{t_p}{2E^{\pm s_p}_{\mathbf{p}}}\frac{t_p E^{\pm s_p}_{\mathbf{p}} - s_p\xi^{\pm s_p}_{\mathbf{p}}}{p_0 - t_p E^{\pm s_p}_{\mathbf{p}}}\Lambda^{-s_p}_{\mathbf{p}}\gamma_0\mathcal{P}_{\rm rg} + \sum_{s_p}\frac{\Lambda^{-s_p}_{\mathbf{p}}\gamma_0\mathcal{P}_{\rm b}}{p_0 + s_p\xi^{\pm s_p}_{\mathbf{p}}}, \tag{7.22}$$

$$\mathbf{F}^\pm_p = {\rm i}\sum_{s_p}\sum_{t_p}\frac{t_p}{2E^{\pm s_p}_{\mathbf{p}}}\frac{\Delta^\pm}{p_0 - t_p E^{\pm s_p}_{\mathbf{p}}}\Lambda^{s_p}_{\mathbf{p}}\gamma_5\tau_2\lambda_2, \tag{7.23}$$

where $s_p, t_p = \pm 1$, $(\Delta^+, \Delta^-) = (\Delta^*, \Delta)$. For the subsequent evaluation of traces in quark-loop diagrams, it is convenient to use this notation with projectors in color space, $\mathcal{P}_{\rm rg} = {\rm diag}(1,1,0)$, $\mathcal{P}_{\rm b} = {\rm diag}(0,0,1)$, and in Dirac space,

$$\Lambda^{\pm}_{\mathbf{p}} = \frac{1}{2}\left[1 \pm \gamma_0 \left(\frac{\boldsymbol{\gamma}\cdot\boldsymbol{p} + \hat{m}}{E_{\mathbf{p}}}\right)\right].$$

The summation over Matsubara frequencies $p_0 = i\omega_n$ is most systematic using the above decomposition into simple poles in the p_0 plane. The poles of the normal propagators $\mathbf{G}^{\pm}$ are given by the gapped dispersion relations for the paired red–green quarks (antiquarks), $E^-_{\mathbf{p}}\,(E^+_{\mathbf{p}})$, and the ungapped dispersions $\xi^-_{\mathbf{p}}\,(\xi^+_{\mathbf{p}})$ for the blue quarks (antiquarks). The anomalous propagators $\mathbf{F}^{\pm}_{\mathbf{p}}$ are only nonvanishing in the 2SC phase when the pair amplitude is nonvanishing. Let us notice explicitly that this procedure has yielded an effective action that includes the fluctuation terms responsible for the excitation of scalar and pseudoscalar mesonic modes. The evaluation of the traces (7.20) can be performed with the result

$$\frac{1}{2}{\rm Tr}\,(S_{\rm MF}\Sigma S_{\rm MF}\Sigma) = (\boldsymbol{\pi}, \sigma)\begin{pmatrix} \Pi_{\pi\pi} & 0 \\ 0 & \Pi_{\sigma\sigma}\end{pmatrix}\begin{pmatrix}\boldsymbol{\pi} \\ \sigma\end{pmatrix}, \tag{7.24}$$

with

$$\Pi_{\sigma\sigma}(q_0,\mathbf{q}) \equiv {\rm Tr}[\mathbf{G}^+_p\mathbf{G}^+_{p+q} + \mathbf{F}^-_p\mathbf{F}^+_{p+q} + \mathbf{G}^-_p\mathbf{G}^-_{p+q} + \mathbf{F}^+_p\mathbf{F}^-_{p+q}], \tag{7.25}$$

$$\begin{aligned}\Pi_{\pi\pi}(q_0,\mathbf{q}) \equiv {} & -{\rm Tr}[\mathbf{G}^+_p(\gamma_5\boldsymbol{\tau})\mathbf{G}^+_{p+q}(\gamma_5\boldsymbol{\tau}) + \mathbf{F}^-_{\mathbf{p}}(\gamma_5\boldsymbol{\tau}^t)\mathbf{F}^+_{p+q}(\gamma_5\boldsymbol{\tau}) \\ & + \mathbf{F}^+_p(\gamma_5\boldsymbol{\tau})\mathbf{F}^-_{p+q}(\gamma_5\boldsymbol{\tau}^t) + \mathbf{G}^-_p(\gamma_5\boldsymbol{\tau}^t)\mathbf{G}^-_{p+q}(\gamma_5\boldsymbol{\tau}^t)].\end{aligned} \tag{7.26}$$

These polarization functions are the key quantities for the investigation of mesonic bound and scattering states in quark matter. In the following, we perform the further evaluation and discussion for the pionic modes, and the σ modes is treated in an analogous way. We start with the evaluation of traces and Matsubara summation.

$$\begin{aligned}\Pi_{\pi\pi}(q_0,\mathbf{q}) = 2\int\frac{{\rm d}^3p}{(2\pi)^3}\sum_{s_p,s_k}\mathcal{T}^+_-(s_p,s_k)\Bigg\{ & \frac{n_{\rm F}(s_p\xi^{s_p}_{\mathbf{p}}) - n_{\rm F}(s_k\xi^{s_k}_{\mathbf{p+q}})}{q_0 - s_k\xi^{s_k}_{\mathbf{p+q}} + s_p\xi^{s_p}_{\mathbf{p}}} \\ & - \frac{n_{\rm F}(s_p\xi^{s_p}_{\mathbf{p}}) - n_{\rm F}(s_k\xi^{s_k}_{\mathbf{p+q}})}{q_0 + s_k\xi^{s_k}_{\mathbf{p+q}} - s_p\xi^{s_p}_{\mathbf{p}}} \\ & + \sum_{t_p,t_k}\frac{t_pt_k}{E^{s_p}_{\mathbf{p}}E^{s_k}_{\mathbf{p+q}}}\frac{n_{\rm F}(t_pE^{s_p}_{\mathbf{p}}) - n_{\rm F}(t_kE^{s_k}_{\mathbf{p+q}})}{q_0 + t_kE^{s_k}_{\mathbf{p+q}} - t_pE^{s_p}_{\mathbf{p}}} \\ & \times\left(t_pt_kE^{s_p}_{\mathbf{p}}E^{s_k}_{\mathbf{p+q}} + s_ps_k\xi^{s_p}_{\mathbf{p}}\xi^{s_k}_{\mathbf{p+q}} - |\Delta|^2\right)\Bigg\},\end{aligned} \tag{7.27}$$

where

$$\mathcal{T}_{-}^{+}(s_p, s_k) = \left(1 + s_p s_k \frac{\mathbf{p} \cdot \mathbf{p} + \mathbf{q} - m^2}{E_{\mathbf{p}} E_{\mathbf{p}+\mathbf{q}}}\right). \tag{7.28}$$

For a pionic mode at rest in the medium ($\mathbf{q} = 0$), this reduces to

$$\begin{aligned}
\Pi_{\pi\pi}(q_0, \mathbf{0}) = 8 \int \frac{\mathrm{d}^3 p}{(2\pi)^3} \Bigg\{ & N(\xi_{\mathbf{p}}^+, \xi_{\mathbf{p}}^-) \left[\frac{1}{q_0 - 2E_{\mathbf{p}}} - \frac{1}{q_0 + 2E_{\mathbf{p}}}\right] \\
& + \left[1 - \frac{\xi_{\mathbf{p}}^+ \xi_{\mathbf{p}}^- + \Delta^2}{E_{\mathbf{p}}^+ E_{\mathbf{p}}^-}\right] M(E_{\mathbf{p}}^+, E_{\mathbf{p}}^-) \\
& \times \left[\frac{1}{q_0 - E_{\mathbf{p}}^+ + E_{\mathbf{p}}^-} - \frac{1}{q_0 + E_{\mathbf{p}}^+ - E_{\mathbf{p}}^-}\right] \\
& + \left[1 + \frac{\xi_{\mathbf{p}}^+ \xi_{\mathbf{p}}^- + \Delta^2}{E_{\mathbf{p}}^+ E_{\mathbf{p}}^-}\right] N(E_{\mathbf{p}}^+, E_{\mathbf{p}}^-) \\
& \times \left[\frac{1}{q_0 + E_{\mathbf{p}}^+ + E_{\mathbf{p}}^-} - \frac{1}{q_0 - E_{\mathbf{p}}^+ - E_{\mathbf{p}}^-}\right] \Bigg\},
\end{aligned} \tag{7.29}$$

where we have introduced the phase space occupation factors $N(x, y) = 1 - n_{\mathrm{F}}(x) - n_{\mathrm{F}}(y)$ (Pauli blocking) and $M(x, y) = n_{\mathrm{F}}(x) - n_{\mathrm{F}}(y)$. For $\mu \neq 0$, this function has three poles from the first terms in each bracket, corresponding to positive energies ($q_0 > 0$). So we need to focus only on these three terms. For $\mu = 0$, the second term vanishes due to the prefactor and we are left with two poles.

We make use of the Dirac identity $\lim_{\eta \to 0} \frac{1}{x + \mathrm{i}\eta} = \mathcal{P}\frac{1}{x} - \mathrm{i}\pi\delta(x)$ to decompose the polarization function into real and imaginary parts after analytical continuation to the complex plane. The imaginary part is straightforwardly integrated after transformation from momentum to energy ω. At the pole, the variables transform as

$$p_\omega = \sqrt{\frac{\omega^4 - 4\omega^2(\mu^2 + \Delta^2)}{4(\omega^2 - 4\mu^2)} - m^2}. \tag{7.30}$$

For $\eta_{\mathrm{D}} < 1$, we know that $\Delta = 0$ if $m \geq \mu$, which includes that $\omega \geq 2\mu$ as this is the relevant threshold. Therefore, the pole is not hidden and we recover the usual $2m$ threshold. For small enough couplings, $\Delta \neq 0$ only if $m < \mu$. Therefore, this pole is not hidden in this case. This reasoning includes that the argument of the square root is strictly positive. The integration borders thus shift $p \in (0, \infty) \to \omega \in (X_{\pm}, \infty)$, where the thresholds are given by $2m$ and

$$X_{\pm} = \sqrt{(m + \mu)^2 + \Delta^2} \pm \sqrt{(m - \mu)^2 + \Delta^2}. \tag{7.31}$$

The pion polarization function in the 2SC phase can thus be decomposed into real and imaginary parts in the following form:

$$
\begin{aligned}
\Pi^{\Delta}_{\pi\pi}(\omega+\mathrm{i}\eta,\mathbf{0}) &= \mathrm{Re}\Pi^{\Delta}_{\pi\pi}(\omega+\mathrm{i}\eta,\mathbf{0}) + \mathrm{i}\mathrm{Im}\Pi^{\Delta}_{\pi\pi}(\omega+\mathrm{i}\eta,\mathbf{0}) \\
&= 8\int\frac{\mathrm{d}^3p}{(2\pi)^3}\left\{N(\xi^+_{\mathbf{p}},\xi^-_{\mathbf{p}})\left[\frac{\mathcal{P}}{\omega-2E_{\mathbf{p}}}-\frac{1}{\omega+2E_{\mathbf{p}}}\right]\right. \\
&\qquad + \left[1-\frac{\xi^+_{\mathbf{p}}\xi^-_{\mathbf{p}}+\Delta^2}{E^+_{\mathbf{p}}E^-_{\mathbf{p}}}\right]M(E^+_{\mathbf{p}},E^-_{\mathbf{p}}) \\
&\qquad \times\left[\frac{\mathcal{P}}{\omega-E^+_{\mathbf{p}}+E^-_{\mathbf{p}}}-\frac{1}{\omega+E^+_{\mathbf{p}}-E^-_{\mathbf{p}}}\right] \\
&\qquad - \left[1+\frac{\xi^+_{\mathbf{p}}\xi^-_{\mathbf{p}}+\Delta^2}{E^+_{\mathbf{p}}E^-_{\mathbf{p}}}\right]N(E^+_{\mathbf{p}},E^-_{\mathbf{p}}) \\
&\qquad \left.\times\left[\frac{\mathcal{P}}{\omega-E^+_{\mathbf{p}}-E^-_{\mathbf{p}}}-\frac{1}{\omega+E^+_{\mathbf{p}}+E^-_{\mathbf{p}}}\right]\right\} \\
&\quad -\mathrm{i}\frac{2}{\pi}\left\{p^0_\omega E_{p^0_\omega}N(\xi^+_{p^0_\omega},\xi^-_{p^0_\omega})\Theta(\omega-2m)+p_\omega E_{p_\omega}\right. \\
&\qquad \times\frac{E^+_{p_\omega}E^-_{p_\omega}-\xi^+_{p_\omega}\xi^-_{p_\omega}-\Delta^2}{\xi^+_{p_\omega}E^-_{p_\omega}-\xi^-_{p_\omega}E^+_{p_\omega}}M(E^+_{p_\omega},E^-_{p_\omega})\Theta(\omega-X_-) \\
&\qquad - p_\omega E_{p_\omega}\frac{E^+_{p_\omega}E^-_{p_\omega}+\xi^+_{p_\omega}\xi^-_{p_\omega}+\Delta^2}{\xi^+_{p_\omega}E^-_{p_\omega}+\xi^-_{p_\omega}E^+_{p_\omega}}N(E^+_{p_\omega},E^-_{p_\omega}) \\
&\qquad \left.\times\,\Theta(\omega-X_+)\right\},
\end{aligned}
\tag{7.32}
$$

where $\mathcal{P}$ denotes the principal value integration, $p^0_\omega = p_\omega\mid_{\Delta=0}=\sqrt{\frac{\omega^2}{4}-m^2}$ and we have made explicit the three thresholds, $2m$ and $X_\pm$, for the occurrence of the corresponding decay processes, giving rise to the partial widths Γ_{2m} and $\Gamma_\pm$, respectively. In the normal phase, this reduces to

$$
\begin{aligned}
\Pi^{0}_{\pi\pi}(\omega+\mathrm{i}\eta,\mathbf{0}) &= \mathrm{Re}\Pi^{0}_{\pi\pi}(\omega+\mathrm{i}\eta,\mathbf{0}) + \mathrm{i}\mathrm{Im}\Pi^{0}_{\pi\pi}(\omega+\mathrm{i}\eta,\mathbf{0}) \\
&= 24\int\frac{\mathrm{d}^3p}{(2\pi)^3}N(\xi^+_{\mathbf{p}},\xi^-_{\mathbf{p}})\left[\frac{\mathcal{P}}{\omega-2E_{\mathbf{p}}}-\frac{1}{\omega+2E_{\mathbf{p}}}\right] \\
&\quad -\mathrm{i}\frac{6}{\pi}p^0_\omega E_{p^0_\omega}N(\xi^+_{p^0_\omega},\xi^-_{p^0_\omega})\Theta(\omega-2m).
\end{aligned}
\tag{7.33}
$$

The analytic properties of the mesonic modes can be analyzed from their spectral function. Here we discuss results for pionic modes with $\mathbf{q}=0$ in the rest frame of the medium

$$
\rho_\pi(\omega+\mathrm{i}\eta,\mathbf{0}) = \frac{8G^2_\sigma\mathrm{Im}\Pi_{\pi\pi}(\omega+\mathrm{i}\eta,\mathbf{0})}{[1-2G_\sigma\mathrm{Re}\Pi_{\pi\pi}(\omega,\mathbf{0})]^2+[2G_\sigma\mathrm{Im}\Pi_{\pi\pi}(\omega+\mathrm{i}\eta,\mathbf{0})]^2}. \tag{7.34}
$$

In the limit of vanishing imaginary part, we recover the spectral function for a "true" (on-shell) bound state

$$\lim_{\mathrm{Im}\Pi_{\pi\pi}\to 0} \rho_\pi(\omega + \mathrm{i}\eta, \mathbf{0}) = 2\pi\delta(1 - 2G_\sigma \mathrm{Re}\Pi_{\pi\pi}(\omega, \mathbf{0})), \tag{7.35}$$

which corresponds to an infinite lifetime of the state and a mass to be found from the pole condition $1 - 2G_\sigma \mathrm{Re}\Pi_{\pi\pi}(m_\pi, \mathbf{0}) = 0$. In Fig. 7.4, we show results for the mass spectrum of pions and sigma-mesons as a function of the temperature for vanishing chemical potential $\mu_\mathrm{B} = 0$ and strong diquark coupling $\eta_\mathrm{D} = 1.0$. As $\Delta = 0$, the only threshold for the imaginary parts of meson decays is $2m$. The σ mass is always above the threshold and therefore this state is unstable in the present model. The pion, however, is a bound state until the critical temperature for the Mott transition $T_\mathrm{Mott} = 212.7\,\mathrm{MeV}$ is reached. For $T > T_\mathrm{Mott}$, the pion becomes unstable for decay into quark–antiquark pairs. As can be seen from the behavior of the spectral function in the lower panel of Fig. 7.4, the pion is still a well-identifyable, long-lived resonance in that case. The detailed analytic behavior of the pion at the Mott transition has been discussed in the context of the NJL model by Hüfner et al. [28], see also the inset of the lower panel of Fig. 7.4. It shows strong similarities with the behavior of bound states of fermionic atoms in traps when their coupling is tuned by exploiting Feshbach resonances in an external magnetic field, see [29]. In the context of RHIC experiments, one has discussed such quasi-bound states as an explanation for the perfect liquid behavior of the sQGP [30].

Next we want to discuss the pionic excitations in the presence of a diquark condensate in the 2SC phase, see Fig. 7.4. We choose $\mu = 320\,\mathrm{MeV}$ and discuss the effect of melting the 2SC diquark condensate by increasing the temperature from $T = 0$ to $T > T_\mathrm{c}$, where $T_\mathrm{c} = 95\,\mathrm{MeV}$ is the critical temperature for the second order transition to the normal quark matter phase. We observe the remarkable fact that the 2SC condensate stabilizes the pion at $T = 0$ as a true bound state, although the pion mass exceeds by far the threshold $2m$. This effect is due to a compensation of gapped and ungapped quark modes and has been discussed before by Ebert et al. [26] for $T = 0$ only. Here we extend this study to the finite temperature case, where the pion is still a very good resonance, but obtains a finite width. At the critical temperature T_c, the normal pion width is restored. But already before $T = T_\mathrm{c}$ is reached, the threshold X_- is reached and the corresponding decay process is opened with a considerable width of $\mathcal{O}(50\,\mathrm{MeV})$. From the pion spectral function in the lower panel of Fig. 7.4, we observe the gap in the excitation spectrum due to the presence of the diquark gap. At $T > T_\mathrm{c}$, a resonance type spectral function with a threshold at $\omega = 2m$ and a resonance peak at $\omega \sim 250\,\mathrm{MeV}$ is obtained.

The discussion of the mesonic modes in the 2SC phase points to a very rich spectrum of excitations, which eventually leads to specific new observable signals of this hypothetical phase. The CBM experiment planned at FAIR

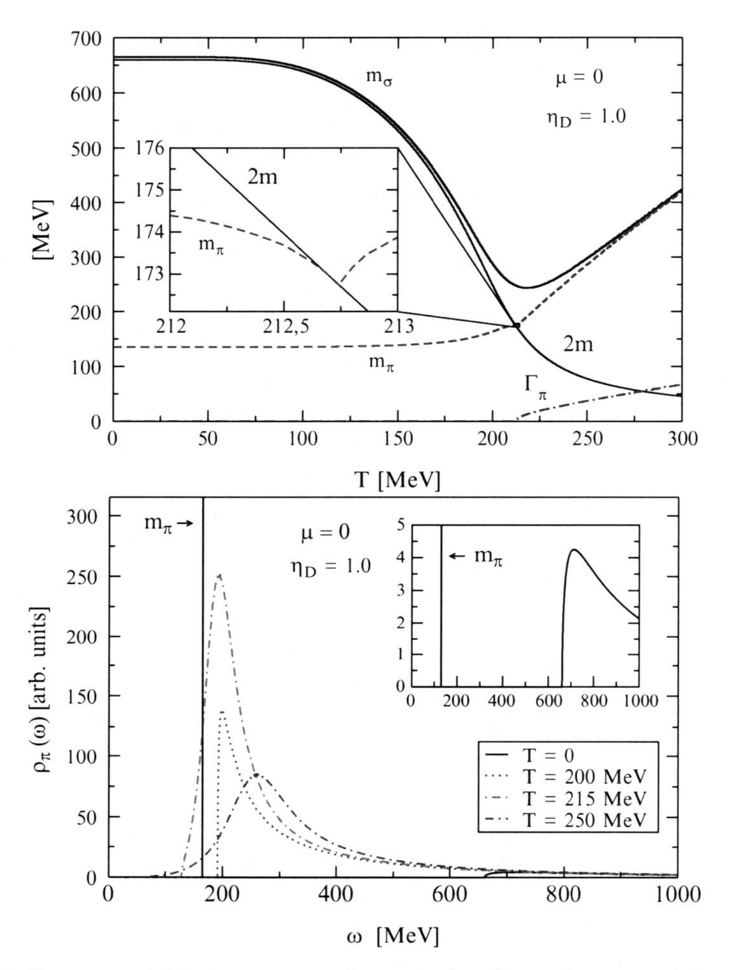

Fig. 7.4. *Upper panel:* Mass spectrum of mesons (π, σ) as a function of the temperature for vanishing chemical potential $\mu_B = 0$ and strong diquark coupling $\eta_D = 1.0$. The threshold $E_{th} = 2m_q$ for Mott dissociation of pions and occurrence of a nonvanishing decay width $\Gamma_\pi = Im\ \Pi_\pi/m_\pi$ is reached at $T_{Mott} = 212.7$ MeV (see *inset*). *Lower panel:* Spectral function for pionic correlations for $\mu_B = 0$ in the vacuum at $T = 0$ (see *inset*) and at different temperatures around the Mott transition. Below T_{Mott}, the bound state (delta function) and the continuum of scattering states are separated by a mass gap. Above T_{Mott}, the spectral function is still sharply peaked, related to a lifetime of pionic correlations of the order of the lifetime of a fireball in heavy-ion collisions (quasi-bound states in the quark plasma)

Darmstadt and the NICA project at JINR Dubna could be capable of creating thermodynamical conditions for the observation of these excitations in the experiment. In view of this discovery potential, we want to outline a few points for the further development of the theoretical approach.

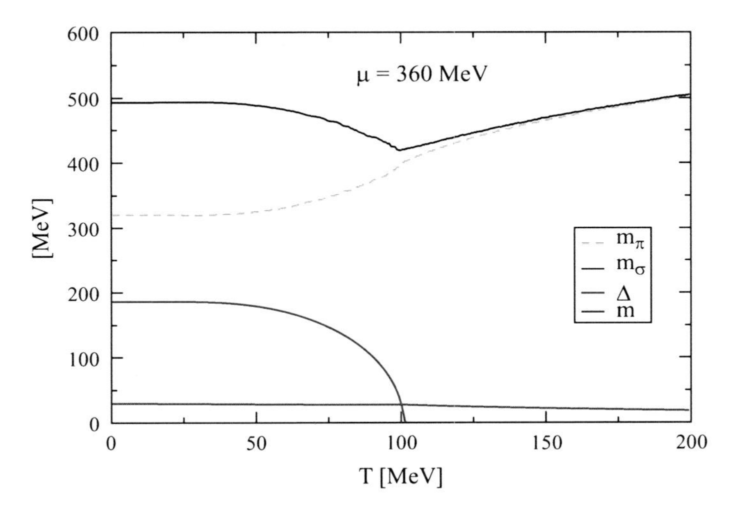

Fig. 7.5. Mass spectrum of mesons (π, σ) as a function of the temperature for finite chemical potential $\mu_B = 360\,\text{MeV}$ and strong diquark coupling $\eta_D = 1.0$ in the 2SC phase

7.3 Further Developments

In this contribution, we have described the first steps into the interesting and very complex physics of the relativistic BEC–BCS crossover theory. As the next steps following this development, some of the approximations can be removed. In particular, one should next:

- Evaluate the full spectrum of diquark states, including their mixing with mesonic channels
- Study the backreaction of the correlations on the meanfield (self-consistent meanfield)
- Include higher orders in the one-fermion-loop approximation (diquark–diquark and diquark–meson interactions)
- study the effect of the color neutrality condition by adjusting color chemical potential(s)
- Study the effect of charge neutrality (gapless superconductivity)
- Evaluate the contribution from diquark–antidiquark annihilation to the photon propagator (Maki–Thompson and Aslamasov–Larkin terms).

In particular, the latter point bears a big potential for applications to the diagnostic of dense quark matter formed, for example, in not too high- energy nucleus–nucleus collisions. The onset of color superconductivity not only changes the spectrum of diquark states (occurrence of the Goldstone bosons) but due to the nonvanishing diquark gap additional terms for the diquark annihilation process into the observable dilepton channel arise, which stem from then nonvanishing anomalous propagator contributions. As the critical temperature for the color superconductivity transition might be as high as 100 MeV, there is a fair chance to observe traces of this transition with the future CBM experiment at FAIR Darmstadt.

7.4 Nuclear Matter

7.4.1 Lagrangian Approach to the Partition Function (NJL vs. Walecka model)

As another generic model system for hot, dense fermionic matter with strong, short-range interactions, we consider nuclear matter described by the NJL model Lagrangian for nucleons [31] with mesonic (σ and ω_0) and nucleon pair (scalar deuteron) interaction channels modeled by a local four-fermion coupling of current–current type. The partition function $\mathcal{Z}$ for this model is given analogously to (7.1) by

$$\mathcal{Z} = \int \mathcal{D}(\mathrm{i}\psi^\dagger)\mathcal{D}(\psi)\, \mathrm{e}^{\int^\beta \mathrm{d}^4x\,(\mathcal{L}-\mu\psi^\dagger\psi)}, \tag{7.36}$$

where in symmetric nuclear matter we have the same chemical potential μ for neutrons (n) and protons (p), which are both described by four-spinors, taken together in the eight-spinor of the nucleon ψ. For details, see [22]. Besides the interaction in the mesonic channels, we want to extend the Walecka model here by $n-p$ pairing interaction, which for simplicity we will assume here to be a Dirac scalar, so that it will generate a scalar $n-p$ bound state (model deuteron) or Cooper pair, depending on the coupling strength. A more realistic model should describe the deuteron as a spin triplet state and therefore by a Dirac vector current, see [32] for details. The Lagrangian assumes the form

$$\mathcal{L} = \mathcal{L}_0 + \mathcal{L}_\mathrm{d} + \mathcal{L}_\mathrm{M}, \tag{7.37}$$

where the different terms are given by

$$\mathcal{L}_0 = \bar{\psi}(\mathrm{i}\gamma_\mu\partial^\mu - m_\mathrm{N})\psi, \tag{7.38}$$

$$\mathcal{L}_\mathrm{M} = g_\sigma\left(\bar{\psi}\psi\right)^2 + g_\omega\left(\bar{\psi}\mathrm{i}\gamma_0\psi\right)^2 \tag{7.39}$$

$$\mathcal{L}_\mathrm{d} = g_\mathrm{d}\{\bar{\psi}\,(\mathrm{i}\gamma_5 C\tau_2)\,\bar{\psi}^\mathrm{T}\}\,\{\psi^\mathrm{T}\,(\mathrm{i}C\gamma_5\tau_2)\,\psi\}\,, \tag{7.40}$$

where g_σ, g_ω, and g_d are the coupling strengths in the scalar meson, vector meson, and deuteron channels, respectively; $C = \mathrm{i}\gamma^2\gamma^0$ is the charge conjugation matrix.

In linearizing these four-fermion interactions in the path integral approach by Hubbard–Stratonovich transformations, we will be able to describe already in the mean-field approximation the liquid–gas phase instability ($\sigma-\omega$ model) and the nuclear superfluidity ($n-p$ pairing).

In principle, the effective Walecka-like Lagrangian (7.37) should be derived from the more fundamental Lagrangian (refLL) after hadronization. In particular, the coupling strengths should be derived that way and a nonlocality of the meson–nucleon coupling (form-factor) emerges from the quark substructure. At present, this approach is not in reach, owing to the fact that the confinement problem has not been solved. It is expected that a unified description of quark and nucleon systems will provide a consistent approach to the properties of dense nuclear matter. At the time being we consider (7.37) as an empirical ansatz.

7.4.2 Hubbard–Stratonovich Transformation: Bosonization

In a complete analogy to the quark matter model, we perform a Hubbard–Stratonovich transformation, introducing the bosonic auxiliary fields $\Delta(\tau, x)$, $\Delta^*(\tau, x)$, $\sigma(\tau, x)$, and $\omega_0(\tau, x)$ so that the partition function of the system becomes

$$\begin{aligned} \mathcal{Z} = &\int \mathcal{D}\Delta^* \mathcal{D}\Delta \mathcal{D}\sigma \mathcal{D}\omega_0 \Big\{ \mathrm{e}^{-\int^\beta \mathrm{d}^4x \left[\frac{\sigma^2}{4g_\sigma} - \frac{\omega_0^2}{4g_\omega} + \frac{|\Delta|^2}{4g_\mathrm{d}} \right]} \\ &\times \int [\mathrm{d}\psi]\,[\mathrm{d}\bar\psi]\, \mathrm{e}^{\int^\beta \mathrm{d}^4x \left(\bar\psi(\mathrm{i}\gamma_\mu\partial^\mu + \mu\gamma_0 - m_N)\psi - \bar\psi(\sigma + \mathrm{i}\gamma_5 \tau\cdot\pi)\psi - \frac{\Delta^*}{2}\psi^\mathrm{T} R\psi - \frac{\Delta}{2}\bar\psi \tilde R \bar\psi^\mathrm{T} \right)} \Big\}. \end{aligned} \tag{7.41}$$

where $R = \mathrm{i}C\gamma_5 \otimes \tau_2$ and $\tilde R = \mathrm{i}\gamma_5 C \otimes \tau_2$. By introducing Nambu–Gorkov spinors

$$\Psi \equiv \frac{1}{\sqrt{2}} \begin{pmatrix} \psi \\ \psi^\mathrm{c} \end{pmatrix}, \quad \bar\Psi \equiv \frac{1}{\sqrt{2}} \left(\bar\psi \bar\psi^\mathrm{c} \right) \tag{7.42}$$

with $\psi^\mathrm{c}(x) \equiv C\bar\psi^\mathrm{T}(x)$, the Lagrangian takes the bilinear form

$$\mathcal{L} = \bar\Psi \begin{pmatrix} \mathrm{i}\gamma_\mu\partial^\mu + \mu^*\gamma_0 - m_\mathrm{N}^* & \mathrm{i}\Delta\gamma_5\tau_2 \\ \mathrm{i}\Delta^*\gamma_5\tau_2 & \mathrm{i}\gamma_\mu\partial^\mu - \mu^*\gamma_0 - m_\mathrm{N}^* \end{pmatrix} \Psi, \tag{7.43}$$

with $m_\mathrm{N}^* = m_\mathrm{N} - \sigma$ and $\mu^* = \mu - \omega_0$. Hence the partition function becomes a Gaussian path integral in the bispinor fields, which can be evaluated and yields the fermion determinant

$$\begin{aligned} \mathcal{Z} &= \int \mathcal{D}\Delta^* \mathcal{D}\Delta \mathcal{D}\sigma \mathcal{D}\omega_0 \mathrm{e}^{-\int^\beta \mathrm{d}^4x \frac{\sigma^2}{4g_\sigma} - \frac{\omega_0^2}{4g_\omega} + \frac{|\Delta|^2}{4g_\mathrm{d}}} \int \mathcal{D}\bar\Psi \mathcal{D}\Psi \mathrm{e}^{\int^\beta \mathrm{d}^4x \bar\Psi [S^{-1}]\Psi} \\ &= \int \mathcal{D}\Delta^* \mathcal{D}\Delta \mathcal{D}\sigma \mathcal{D}\omega_0 \mathrm{e}^{-\int^\beta \mathrm{d}^4x \frac{\sigma^2}{4g_\sigma} - \frac{\omega_0^2}{4g_\omega} + \frac{|\Delta|^2}{4g_\mathrm{d}}} \cdot \mathrm{Det}[S^{-1}], \end{aligned} \tag{7.44}$$

where the inverse bispinor propagator is a matrix in Nambu–Gorkov-, Dirac-, color-, and flavor space, which after Fourier transformation reads

$$S^{-1} = \begin{pmatrix} (\mathrm{i}\omega_n + \mu^*)\gamma_0 - m_\mathrm{N}^* - \mathrm{i}\gamma\mathbf{p} & \mathrm{i}\Delta\gamma_5\tau_2 \\ \mathrm{i}\Delta^*\gamma_5\tau_2 & (\mathrm{i}\omega_n - \mu^*)\gamma_0 - m_\mathrm{N}^* - \mathrm{i}\gamma\mathbf{p} \end{pmatrix}. \tag{7.45}$$

In complete analogy to the quark matter model, we could derive a very compact, bosonized form of the partition function, which is an exact transformation, now formulated in terms of collective, bosonic fields. The subsequent steps are to be carried out in the same manner as in the previous section. We carry on with an expansion of the action functional around stationary, mean field solutions to obtain results in the nonperturbative regime. In performing this expansion, we may factorize the partition function into mean field (MF), Gaussian fluctuation (Gauss), and residual (res) contributions

$$Z(\mu, T) \equiv \mathrm{e}^{-\beta\Omega(\mu,T)} = Z_{\mathrm{MF}}(\mu, T) Z_{\mathrm{Gauss}}(\mu, T) Z_{\mathrm{res}}(\mu, T).$$

In the following, we discuss the physical content of these approximations.

7.4.3 Mean-Field Approximation: Order Parameters and EoS

In thermodynamical equilibrium, the mean field values satisfy the stationarity condition of the minimal thermodynamical potential $\Omega_{\mathrm{MF}} \equiv -\frac{1}{\beta V} \ln \mathcal{Z}_{\mathrm{MF}}$, that is,

$$\frac{\partial \Omega_{\mathrm{MF}}}{\partial \sigma_{\mathrm{MF}}} = \frac{\partial \Omega_{\mathrm{MF}}}{\partial \omega_{\mathrm{MF}}} = \frac{\partial \Omega_{\mathrm{MF}}}{\partial \Delta_{\mathrm{MF}}} = 0, \tag{7.46}$$

equivalent to the fulfillment of the self-consistency equations $\sigma_{\mathrm{MF}} = -4g_\sigma \mathrm{Tr}\,(S_{\mathrm{MF}}) \equiv m_{\mathrm{N}} - m_{\mathrm{N}}^*$, $\omega_{0,\mathrm{MF}} = -4\mathrm{i}g_\omega \mathrm{Tr}\,(\gamma_0 S_{\mathrm{MF}}) = \mu - \mu^*$ and the gap equation $\Delta_{\mathrm{MF}} = 4g_{\mathrm{d}} \mathrm{Tr}\,(\gamma_5 \tau_2 S_{\mathrm{MF}}) = \Delta$, together with the stability criterion that the determinant of the curvature matrix formed by the second derivatives is positive. After the evaluation of the traces in the internal spaces and the sum over the Matsubara frequencies, one gets

$$\begin{aligned}
\Omega_{\mathrm{MF}} &= -\frac{1}{\beta V} \ln \mathcal{Z}_{\mathrm{MF}} = \frac{(m_{\mathrm{N}} - m_{\mathrm{N}}^*)^2}{4g_\sigma} - \frac{(\mu - \mu^*)^2}{4g_\omega} + \frac{|\Delta|^2}{4g_{\mathrm{d}}} - \frac{1}{\beta V} \mathrm{Tr}\left(\ln \beta S_{\mathrm{MF}}^{-1}\right) \\
&= \frac{(m_{\mathrm{N}} - m_{\mathrm{N}}^*)^2}{4g_\sigma} - \frac{(\mu - \mu^*)^2}{4g_\omega} + \frac{|\Delta|^2}{4g_{\mathrm{d}}} - 2\int \frac{\mathrm{d}^3 p}{(2\pi)^3} \Big[E_{\mathbf{p}}^+ + E_{\mathbf{p}}^- \\
&\quad + 2T \ln(1 + \mathrm{e}^{-\beta E_{\mathbf{p}}^+}) + 2T \ln(1 + \mathrm{e}^{-\beta E_{\mathbf{p}}^-}) \Big],
\end{aligned} \tag{7.47}$$

where we have defined the particle dispersion relation $E_{\mathbf{p}}^{\pm} = \sqrt{\left(\xi_{\mathbf{p}}^{\pm}\right)^2 + \Delta_{\mathrm{MF}}^2}$ with $\xi_{\mathbf{p}}^{\pm} = E_{\mathbf{p}} \pm \mu^*$, $E_{\mathbf{p}} = \sqrt{(m_{\mathrm{N}}^*)^2 + \mathbf{p}^2}$. The integral over the zero-point energies in (7.47) is divergent and has to be regularized. Here we drop these terms and understand the vacuum contribution to the σ_{MF} to be absorbed into the definition of the bare nucleon mass m_{N}.

From (7.47) with (7.46), we obtain the self-consistency equations for the order parameters σ_{MF}, ω_{MF}, and Δ_{MF}, which have to be solved self-consistently,

$$m_{\mathrm{N}} - m_{\mathrm{N}}^* = 8g_\sigma \int \frac{\mathrm{d}^3p}{(2\pi)^3} \frac{m_{\mathrm{N}}}{E_{\mathbf{p}}} \left[n_{\mathrm{F}}(E_{\mathbf{p}}^-) \frac{\xi_{\mathbf{p}}^-}{E_{\mathbf{p}}^-} + n_{\mathrm{F}}(E_{\mathbf{p}}^+) \frac{\xi_{\mathbf{p}}^+}{E_{\mathbf{p}}^+} \right], \quad (7.48)$$

$$\mu - \mu^* = 8g_\omega m_{\mathrm{N}} \int \frac{\mathrm{d}^3p}{(2\pi)^3} \left[n_{\mathrm{F}}(E_{\mathbf{p}}^-) \frac{\xi_{\mathbf{p}}^-}{E_{\mathbf{p}}^-} - n_{\mathrm{F}}(E_{\mathbf{p}}^+) \frac{\xi_{\mathbf{p}}^+}{E_{\mathbf{p}}^+} \right], \quad (7.49)$$

$$\Delta_{\mathrm{MF}} = 8g_{\mathrm{d}} \Delta_{\mathrm{MF}} \int \frac{\mathrm{d}^3p}{(2\pi)^3} \left[\frac{n_{\mathrm{F}}(E_{\mathbf{p}}^-)}{E_{\mathbf{p}}^-} + \frac{n_{\mathrm{F}}(E_{\mathbf{p}}^+)}{E_{\mathbf{p}}^+} \right], \quad (7.50)$$

with the Fermi distribution function $n_{\mathrm{F}}(E) = (1 + \mathrm{e}^{\beta E})^{-1}$. For zero temperature, the self-consistency equations take the simple form

$$m_{\mathrm{N}} - m_{\mathrm{N}}^* = 8g_\sigma \int \frac{\mathrm{d}^3p}{(2\pi)^3} \frac{m_{\mathrm{N}}}{E_{\mathbf{p}}} \Theta(\mu^* - E_{\mathbf{p}}), \quad (7.51)$$

$$\mu - \mu^* = 8g_\omega m_{\mathrm{N}} \int \frac{\mathrm{d}^3p}{(2\pi)^3} \Theta(\mu^* - E_{\mathbf{p}}). \quad (7.52)$$

In the limit for vanishing gap, $\Delta = 0$, these equations coincide with the ones derived for the Walecka model in the mean-field approximation [22]. Because of this strong relation to the Walecka model, we can identify the corresponding parameters of the model as

$$g_i = \frac{g_{i,W}^2}{2m_i^2}, \qquad i = \sigma, \omega,$$

where the index W denotes the Walecka model parameters: $m_{\mathrm{N}} = 939\,\mathrm{MeV}$, $m_\sigma = 550\,\mathrm{MeV}$, $m_\omega = 783\,\mathrm{MeV}$, $g_{\sigma,W} = 10.3$, and $g_{\omega,W} = 12.7$ [33].

Thus, this approach represents one possibility to introduce pairing into the problem in complete analogy to the quark matter case. However, while the Walecka model itself is a renormalizable theory, the NJL model is known to be non-renormalizable. Our approach includes the phase transition automatically, by the condition of the global minimum of the thermodynamical potential, so that no explicit Maxwell construction has to be done. As g_{d} is a free parameter of our model, we determine it by fitting the deuteron binding energy. Solutions of the gap equations for the effective mass m_{N}^* and for the renormalized chemical potential μ^* for different temperatures as a function of the chemical potential are shown in Fig. 7.6.

The jump in the order parameters σ and ω_0 indicates a first order phase transition, which for increasing temperatures gets less pronounced and vanishes at the critical temperature T_{c}, which in our simple model is approximately $T_{\mathrm{c}} \approx 20\,\mathrm{MeV}$.

Reinserting the meanfield values of these order parameters into the thermodynamical potential (7.47) results in the pressure as a function of μ for given T (isotherms). The corresponding baryon density can be obtained by derivation with respect to the chemical potential $n = \frac{\partial \Omega}{\partial \mu}$. In Fig. 7.7, we display the behavior of $p(\rho, T)$, the EoS for the present nuclear matter model. There we recognize the instability region corresponding to the coexistence of

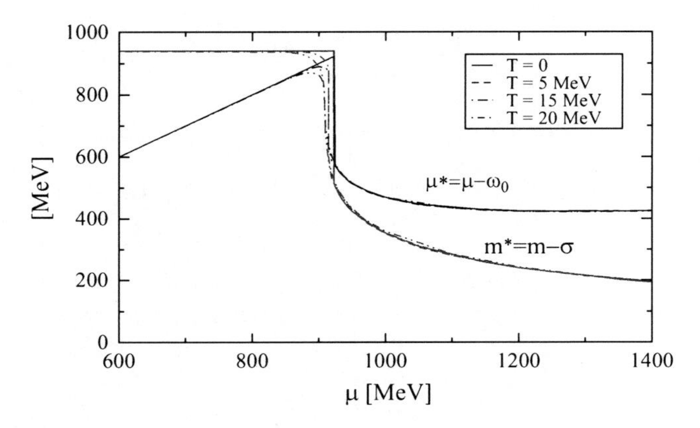

Fig. 7.6. Order parameters as a function of the chemical potential μ for several values of temperature for the case of vanishing pairing gap $\Delta = 0$

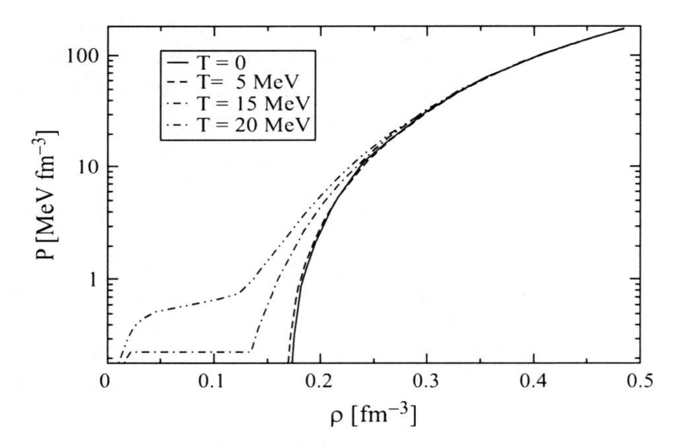

Fig. 7.7. Equation of state for different temperatures. The gas–liquid phase transition is evident from the jump in the density

liquid and gas phases. More details concerning the field theoretical treatment of the nuclear liquid–gas phase transition within the NJL-type Walecka model can be found, for example, in [33, 34].

7.4.4 Discussion

In analogy to the quark matter case, we have suggested to explore the EoS with a Lagrangian approach using the basic channels of attraction (σ) and repulsion (ω_0) in the mesonic channels and pairing (Δ) in the n–p interaction channel. As improvement would be the generalization to nonlocal interactions by, for example, using form-factors of a separable interaction. In this way, a modeling of scattering phase shifts in a dense medium would be possible [11].

Further interaction channels can be included, as also the scalar channel can be modeled in a nonlinear way.

The mesonic and nucleon pair fields have been introduced using the Hubbard–Stratonovich transformation, evaluated in the mean-field approximation, where contact to the Walecka model is established. The new aspect is in the inclusion of the pairing channel, which goes beyond the Walecka model.

In analogy to the situation for the case of the chiral and color superconducting phase transitions in quark matter, which have been discussed above, we can describe the liquid–gas phase transition in the nuclear matter EoS from the behavior of the thermodynamical potential as a function of the order parameter fields.

7.5 Conclusions

We have derived a field theoretical approach to the EoS and the correlations in dense Fermi systems for the example of quark matter and nuclear matter. Further developments are foreseen to provide a unified approach to nuclear matter in terms of quark degrees of freedom, where nucleons and mesons appear as bound or scattering states. This physical picture would allow to discuss the quark–hadron transition due to the change of the order parameters (σ, ω_0, Δ) characterizing the phases of quark/nuclear matter together with the aspect of quark bound state dissociation (delocalization) in analogy to the Mott–Anderson transition. In comparison to earlier nonrelativistic approaches as in [35,36], in the present approach the chiral symmetry restoration and color superconductivity can be implemented in a consistent way. First steps in this direction have been done in [37–39].

The elucidation of the role of nuclear pairing for the self-consistent meanfield EoS within the present approach is one of the frontiers of present research. The coupled set of equations of motion that has been given in the nonrelativistic T-matrix approach in [40] is formulated here within a field theoretic treatment. Within the Gaussian approximation for the pairing fluctuations, the formation of clusters (deuterons) and their dissociation (Mott effect) can be given also below the critical temperature for pair condensation, where this effect is addressed as BEC–BCS crossover in nuclear matter [12].

References

1. Q. Chen, J. Stajic, K. Levin, Low Temp. Phys. **32**, 406 (2006)
2. M. Greiner, C.A. Regal, D.S. Jin Nature **426**, 537 (2003)
3. M.W. Zwierlein, C.A. Stan, C.H. Schunck, S.M. Raupach, S. Gupta, Z. Hadzibabic, W. Ketterle, Phys. Rev. Lett. **91**, 250401 (2003)
4. M.W. Zwierlein, J.R. Abo-Shaeer, A. Schirotzek, C.H. Schunck, W. Ketterle, Nature **435**, 1047 (2003)

5. M. Greiner, O. Mandel, T. Rom, A. Altmeyer, A. Widera, T.W. Hänsch I. Bloch, Physica B **329**, 11 (2003)
6. E. Calzetta, B.L. Hu, A.M. Rey, Phys. Rev. A **73**, 023610 (2006)
7. N. Mott, Rev. Mod. Phys. **40**, 677 (1968)
8. A. Sedrakian, J.W. Clark, M. Alford, eds. *Pairing in fermionic system*, (World Scientific Publications, Singapore, 2006)
9. F.X. Bronold, H. Fehske, Phys. Rev. B **74**, 165107 (2006)
10. R. Redmer, B. Holst, H. Juranek, N. Nettelmann, V. Schwarz, J. Phys. A **39**, 4479 (2006)
11. M. Schmidt, G. Röpke, H. Schulz, Ann. Phys. **202**, 57 (1990)
12. H. Stein, A. Schnell, T. Alm, G. Röpke, Z. Phys. A **351**, 295 (1995)
13. A. Schnell, G. Röpke, P. Schuck, Phys. Rev. Lett. **83**, 1926 (1999)
14. M. Kitazawa, T. Koide, T. Kunihiro, Y. Nemoto, Phys. Rev. D **65**, 091504 (2002)
15. M. Kitazawa, T. Koide, T. Kunihiro, Y. Nemoto, Phys. Rev. D **70**, 056003 (2004)
16. D. Blaschke, D. Ebert, K.G. Klimenko, M.K. Volkov, V.L. Yudichev, Phys. Rev. D **70**, 014006 (2004)
17. D. Blaschke, S. Fredriksson, H. Grigorian, A.M. Öztas, F. Sandin, Phys. Rev. D **72**, 065020 (2005)
18. H. Abuki, Nucl. Phys. A **791**, 117 (2007)
19. J. Deng, A. Schmitt, Q. Wang, Phys. Rev. D **76**, 034013 (2007)
20. G. Sun, L. He, P. Zhuang, Phys. Rev. D **75**, 096004 (2007)
21. E.V. Shuryak, arXiv:nucl-th/0606046
22. J. Kapusta ed. *Finite-temperature Field Theory*, (Cambridge University Press, Cambridge, 1989), p. 26
23. M. Buballa, Phys. Rep. **407**, 205 (2005)
24. H. Grigorian, Phys. Part. Nucl. Lett. **4**, 223 (2007)
25. H. Kleinert, Fortschr. Phys. **26**, 565 (1978)
26. D. Ebert, K.K. Klimenko, V.L. Yudichev, Phys. Rev. C **72**, 015201 (2005)
27. D. Zablocki, D. Blaschke, R. Anglani, AIP conf. Proc. **1038**, 159 (2008)
27a. D. Blaschke, D. Zablocki, Phys. Part. Nucl. **39**, 1010 (2008)
28. J. Hüfner, S.P. Klevansky, P. Rehberg, Nucl. Phys. A **606**, 260 (1996)
29. V. Gurarie, L. Radzihovsky, Ann. Phys. **322**, 2 (2007)
30. E.V. Shuryak, I. Zahed, Phys. Rev. D **70**, 054507 (2004)
31. Y. Nambu, G. Jona-Lasinio, Phys. Rev. **122**, 345 (1961); Phys. Rev. **124**, 246 (1961)
32. A.N. Ivanov, H. Oberhummer, N.I. Troitskaya, M. Faber, Eur. Phys. J. A **7**, 519 (2000)
33. M. Buballa, Nucl. Phys. A **611**, 393 (1996) [arXiv:nucl-th/9609044]
34. H. Reinhardt, H. Schulz, Nucl. Phys. A **432**, 630 (1985)
35. C.J. Horowitz, E.J. Moniz, J.W. Negele, Phys. Rev. D **31**, 1689 (1985)
36. G. Röpke, D. Blaschke, H. Schulz, Phys. Rev. D **34**, 3499 (1986)
37. W. Bentz, A.W. Thomas, Nucl. Phys. A **696**, 138 (2001)
38. R. Huguet, J.C. Caillon, J. Labarsouque, Nucl. Phys. A **781**, 448 (2007)
39. A.H. Rezaeian, H.J. Pirner, Nucl. Phys. A **769**, 35 (2006)
40. A. Sedrakian, Prog. Part. Nucl. Phys. **58**, 168 (2007)

Index